ODYSSEY of the AGES ...

Human Journeys in Time and Space ... at 370,000 Times the Speed of Light!

By
Robert B Cronkhite

Other Books by Robert B Cronkhite

Non-fiction
Hypersphere: A Journey at the Speed of Geometry

Non-fiction
Architecture for a Third American Republic

Fiction
Ballad 'O Stone ... The Trail of a Lonesome Fugitive

With such,
now we are able to better see our place
as in a theater of time and space
when and where it can be said in the Hebraic,
'ahava', or love ...

"The smile of a baby,
and the heat of 400 million suns!"

Odyssey of the Ages

Prologue 1

Part I **120,000 BC to 120,000 AD ...**
            Ancient History from a Future Time

Chapters

Midlogue   149

Part II 2010 AD To 1895 AD ...
  The Personal Temporal Effects
  of Hypergeometric Mechanics

Chapters

Prologue

These are the renderings of human participation in practical mechanism, dramatic or experiential, within the contexts of Hypergeometric Mechanics. The theoretical is no longer abstract and esoteric, but as machine. What if one actually pursues such ventures without the limit of lingering doubts of attainment, and dares to voyage into what was before only glimpsed? It is the proposal of new worlds, so far in time and space, yet only millimeters and less away ... the function of such processes, so powerful that for our humanity it is bewildering and astounding and wonderful all at the same time ... at the speed of geometry 370,000 times the speed of light!

Here within, we have two parts of such an epic of grand proportions of time, deep time, and space, deep space, and the parallel human attempt to attain only a passing grasp of a grander reality in which its own history has, is and will unfold. It is with daring that the attempts of those herein at some point in their lives actually achieve, revealing humbled humanity to its place upon the cosmic stage.

In 2010 AD within the quiet and forgotten concepts of those overlooked by academia, national interest of governments and the elitists, a few discovered an immense release into other times and other worlds. Now emerges in script a composition of grandeur that puts into context humankind's odyssey amongst the stars in our galaxy and in our time within the rotations of the eons.

Extrapolations swiftly minimize our arrogance and at the same time allow us to peer at moments of those things so long before and yet to come. We are better able to perceive our position within much grander structures.

Enter now into vistas that oft begin with quiet moments and rare things, to within swift effect powerful perspectives that can melt the heart and still the mind of those daring, or without suspect, find themselves in new worlds and times!

Yes, it is also said, 'Oft times the most profound begins with the most subtle'!

Part 1

120,000 BC to 120,000 AD …
Ancient History from a Future Time

Chapter 1

History of a Lost World

It has been a pathetic history of such a lost and tragic race. Wars and rumors of wars, continual disappointment and repeated unfulfilled promises … it would be ages before reflection gave way to insight.

Decades had passed until the awaited signal from humanity's first interstellar endeavor to the nearby stars of Alpha Centauri utilizing x-ray pulsar navigation was successfully received. Finally during the last few weeks the signals have been pouring in from the red laser onboard, bringing to Earth images and information of an older, Earth-like planet of deserts and remains of an oceanic past. This orbiting relic in the habitable zone of the aging Alpha Centauri A sun had such details to share.

In the late twenty-first century it was said that there were ruins similar to a more ancient Earth in such constructs. It was even discussed in Biblical debates of a time when many such grand buildings existed, fortuitous that a few remained to present day. Their builders were perhaps giants others had speculated. Even Enoch of old had seen them. Where did they come from?

It has been for millennia that our own sun and the suns of Alpha Centauri closely co-orbited in this part of our Milky Way Galaxy. Was it from here that those giants had come that had interfered with humanity as some have wondered in Genesis chapter six?

In these images I remembered seeing those apparent, then more pronounced large ruins, having been built by someone or something of quite great size and strength. These were so similar to the few that remain to this day in Egypt, North, Central and South America, and even in remote parts of Asia. These artifacts of such a long past prior intelligence were quite large and heavy in appearance. Simple and massive had to be their construction, and so implied larger articulations of their labor. It also was of the implication of a giant race of creatures with a pseudo-human design of like proportions.

Are there other lost worlds with lost histories? It was St. Augustine who had said, "My God is the maker of worlds" and I add 'He is also the destroyer of such!' Are we possibly seeing where the giants of Genesis and its ancient history came from?

The ancients had called them the great 'builders' who of ages gone by built the pyramids and perhaps other mysterious large structures scattered across our Earth. They had been mentioned in the Book of Enoch, the Book of Jasher and specifically in the sixth chapter of the Book of Genesis, as well as in ancient secular accounts. These 'men' of renown, these giants of long ago, building such large and imposing structures so well in their designs connected to the stars in the night sky.

With the extrapolations that Earth's sun and the Alpha Centauri System of suns have, in enough recent and ancient terrestrial history, danced through our Galaxy for an extremely long time. With these stars by analysis being at least a billion years older than our sun, we are so easily left to consider that amongst the stars there are some form of 'fallen angels'... non-human intelligence and technology. That we, while in our more primitive state, influenced our own species as recorded, considered this but a myth in our supposed enlightened age.

It was in the latter twentieth and early twenty-first centuries that robotic probes would free our observations and measurements to ground level. Ground truth as it is defined, upon the surface of our moon, Mars and Titan. The atmospheres and surfaces of the other planets were also closely reckoned. As time has passed the other planets of other solar systems in the great creation were also coming of age in observation and measure. Alpha Centauri though exceedingly remote in distance is a compatriot neighbor in close proximity to human history. Because of the forest the trees were not seen. Yet even if postulated, are these perhaps siblings of our sun? Fascinating does the consideration unfold so logically with credible answers to some of the mysteries of our ancient past.

An argumental ploy of someone trying to convince others to his position is usually really a one way conversation. This tactic in academic and social banters is speaking, but not listening ... no longer conversation, but a monologue of their own opinions. Whether in the prehistoric past Earth was visited by some advanced, non-human intelligence or during our later, unrecorded history or not, we question where they might have come from and what was their origin? Was it nearby or far away? Could these non-human entities of intelligence and technology really be the fallen angels of the Biblical account? Even dragons are another such thing of the parlor debates as one sips with the others some brandy and begin to go tangent in the thoughts being discussed. Some elements of reality and fact are attributed to mysteries of myth and legend. Similarly even the roundabout game of 'gossip', with the first person whispering a certain phrase to the second person passing on from mouth to ear amongst the participants, with the final recitation an embellishment of the original seed, a more elaborate flower.

Presently our own technology has developed into the ability to circumspect the nearby planets and the Alpha Centauri triple star system. Finally we were able to inspect them as we had our own solar system, when decades ago we had actually been amongst them. That a very fine situation to repeat what may have happened to our own sun's worlds ages ago.

The raw image data from the blue laser onboard continued and any interstellar dust, then from the red laser came trickling in. It would have to be processed and cleaned up so to speak to enable the finer details to emerge. 4.32 light years was the average distance for this transfer, thus everything coming in was delayed by 4.32 light years; as such this was not a problem for the greater span of history's intrigues over the last thousands or more years. From these long traveling photons came the beginnings of new interpretations of old and yet partially known things.

When we finally had gotten the latest of the images, there appeared on the desert world of the ancient Earth-like planet, ruins. Large scale ruins, the builders of such very large and strong as well. Perhaps there were two classes ... the designers and the builders, one of intelligence and the other of brute strength.

What if such non-human, technological intelligence had expanded from its proper domain and thusly at some far prior time, infiltrated upon our own early history? Perhaps it was these fallen ones who had come from the sky suddenly, but with ample time to influence humanity so long ago. God had made them too, but they purposed to leave their place to intrude upon those in God's image, such as the Genesis account and other early human accounts of this Earth of ours imply.

There has been such a compilation of information coming in, that what was more astounding was that the past was being revealed in the future. While that premise was being confirmed, it also assured that the future reveals the past.

What nearby discovery would be revealed not only upon Earth, but from Earth. As human pre-history and our earliest recorded history are of our time context, our solar system of the moon and planets are our space context. Amongst the mysteries of archaeology, anthropology and paleontology are perhaps artifacts of these giant-builders.

Often, I had gone over the archives of the lunar and Martian images and wondered about some of the yet to be revealed intrigues ... caves found on the northern floor of the great crater Copernicus on the near side of our natural satellite. Some odd formations vexing natural geologic explanation in other craters and on Mars there were similar strange, unaccountable structures noticed.

Transient, lunar articulations from since the eighteenth century, have provided no answers to the secrets of the fourth planet from the Sun, it being farther away. Yet even with the advent of robotic space probes, Mars continued to raise questions. Some would be solved and others remain a mystery. Back on the Moon, it had been easier, yet no less unnerving to consider implications of what was nearby and more accessible to our scrutiny. Would what was now being seen 4.32 light years away, appear so strange and yet so familiar?

On a couple of the return flights to the Moon, the astronauts and a cosmonaut had investigated the caves of Copernicus. It was discovered to be a crater upon a lava flood plain in the Oceanus Procellarum Mare and its own impact had brought up hot lava material which layered and cooled after the initial eruption of some 800 million years ago. As the layers cooled some issued forth gas creating voids that left pockets of empty space under the new surface. In time as more cooling occurred and as the stresses of orbiting Earth with a tidal lock upon our moon, these tidal movements caused some sections to collapse and open forming these caves in Copernicus, as well as other areas of the Moon. Domes would also arise filled with internal gases; some would escape in time leaving empty but closed chambers.

All in all, these caverns on our moon and also on Mars were safe places for our spacefarers to dwell in protection from solar and galactic cosmic rays and other radiations. It was a logical and truly primitive response that all life seeks protection in such environments. Also the giants, as any life form also probably sought such protection. This intrigue would propagate from what some have considered remains of tools, pottery, stone work and various odds and ends.

Even small, metal spheres similar to ball bearings were found, much like those found in a deep gold mine shaft in South Africa decades ago. In one cave the remains of what at first seems a 'natural' nuclear pile or reactor was recently discovered. Akin to the Oko natural nuclear reactor that was found in Africa quite a long time ago, both still quite preserved.

All this strange yet familiar, not only within our own solar system, but now in the Alpha Centauri System suggested some connection. Waiting for the results from the advanced landers would be very much welcome. Perhaps our human past was far more fascinating and far more complicated than we cared to believe.

Clockwork universe ... better clockwork superuniverse? Despite if one agrees or not, interpretations of time in Genesis, God's Universe, better God's Super-Universe has been working for billions and possibly more years. If the conduit of free-will within the conduit of determinism were not so, it would not work. There is the variability of free-will within the deterministic trends of the greater cylinder of the inherent world line.

God's Superuniverse and its human history within, an account that would find its script quite circular not only upon a spherical globe, but within the spherical temporal geometry as well. The past revealing future; the future revealing past!

Chapter 2

Ruins of a Far More Distant Place

It took months for the information that kept coming in to be processed and analyzed. Were these the architectural remains of the fallen angels of old and Biblical renown? Could this make sense in our time, of eons ago when and where human origins lie? I had to pursue these streams of new, revealing information that now allowed me to better investigate what had been hidden in the past as myth and legend. This was verification of what many others considered true, but until now without rendering beyond our past limited perspectives.

Speculations following my ponderings were my recurring investigations. Did we advance that far as to have originated here and go there; or did we originate there and come here? Perhaps another question, one that ponders parallel origins but in nearness of Galactic locale, could we have been created here; and they, those dark angel-giants, there?

If one considers a bevy of bees or a colony of ants do they not tend to radiate from a greater population to farther trends of proper environment to spread their kind and to feed? If we look upon a greater scale, humanity seems to also spread from certain concentrations of population to farther, more suitable environs.

In prehistory or at least history lost or only partially remaining, could we have interacted with our nearby stellar environment, within our Galactic ecosystem, the Milky Way? If certain stars that are similar, benefit life and social beings (human or non-human, including their technology), cohabitated close to Galactic orbits for long periods of time, sometimes interchanging dust, comets and other particulates, then could they have been sharing life as well?

Consider life and its resultant possibilities of technology, then could such life be enabled enough to cross the distances of nearby stellar realms to explore and even colonize and populate greater environs?

What in our written history is lacking in such a probability? If given enough time and space, more and more is possible, even based on God's probabilistic mathematics. This would bring into certainty and conviction that which was Biblically mentioned and that which has been implied by some archaeologically and anthropologically. Perhaps this is not only in our Galaxy, but in others; though rare, still occurring.

Again, given enough time and enough space, the rare and extraordinary is more likely to happen. Our God is limited by no human being's theological box or any other contrived limit. Our Maker of Worlds has as His 'ocean' the Superuniverse and that is far more than we can fathom. We are simply sampling such grander things of His Greater Ways.

Perchance we have discovered further of this grander story, as per our own past and beginnings. Such ruins were so surprisingly reminiscent of the period of human history before the Great Flood of Earth's ancient past. Now to try to put all this together meant that more probes would have to be sent, more expeditions to Earth's early history of humankind's civilizations, even to the time of hunter and gatherer, and agriculture to village. Somewhere in this progression did someone or something intervene?

Not only archaeology, but anthropology is relevant. Even Paleontology is to be considered, for origins start out from the progressions of other origins. Origins promulgate from one point to another, they are the beginnings from the endings and are eternal, it seems.

We usually pick a point in time or place to call a 'beginning' and another an 'ending'; yet to focus down upon that point like a demarcation with more resolution shows us more of the spectrum's details. The deeper we go to measure, the more we find, not less. At some place and time even the Inertial Geometric, where we only find mass, momentum and inertia, tends to fade into the massless and momentum-less and inertia-less, at 370,000 times the speed of light. Here, when time as we would observe and measure it would be in reverse and only past. Future by time dilation is inertial; it follows the below and near light speed where the Lorentzian effects give us time dilation forward in time. Go too fast and the entanglement, superpositioning and wormholing becomes the norm. Go fast enough and far enough, and the past is the future. And go nearer the speed of light, keeping one's increasing inertia, and the future is the past. Our past is as fantastic as our future. 'Now' passes by in the resonance of all the inclusive kinetics within the local curvature of space and time, at one second per second.

It is written: "There is nothing new under the Sun", and what of the more distant suns ... what more could be said? How far have we then fallen from our greater state? What was it like before the fallen angels, those giants?

Going back far enough to the interglacial period, that previous warm period before the last Ice Age, caused me to contemplate what Florida might have looked like ... simply some islands with sand dunes and sandy beaches? Then only its above water environment would be these presently highest hills. And with all of the sea life, their shells would be the fossils for our future museums. Our own Ice Age would arrive perhaps a million years later then. From Florida to the Bahamas it would mean the other extreme. Now with ice holding the waters of our oceans in deep freeze, these same areas were far, extended grasslands with migrating megafauna and megaflora. Here would be the arrival in this period around 120,000 BC of the visitors from Alpha Centauri, these strangers who interfered with our ancestors' hunting and gathering to follow such grand migrations to survive. Yes, I am so trying to extrapolate the continuing stream of bits and pieces of our earliest history.

I feel like an historian, archaeologist and anthropologist all rolled into one. But the implications are astounding. Here the future is more and more revealing the past. It is taking this specific Alpha Centauri Probe to unravel and offer pieces to complete the past puzzle of our origins.

Chapter 3

Speculations upon Eden

Let's reconsider that the Alpha Centauri System is not only co-rotating, but possibly even gravitationally bound at quite an extenuation of the local bounds of our Sun's Solar System. That being said, it would be far more likely that not only were the Alpha Centauri stars from the same birth cloud as our own sun, but quite possibly have a joint history in their shared travel in our Milky Way Galaxy. Also a higher percentage of exchange between them of dust and gases, inorganic and organic molecules, including biological entities of at least the microscopic scale could have occurred. Extrapolating to more advanced creativities beyond, many futilely put God in their self-satisfying boxes, that He in His Unlimited Power could have raised others of society and intelligence. Could the rebellious dark angels have come from here alone, or from here and other farther places as well? Are perhaps the greater bounds of celestial space, theologically the second heaven, abodes of good and bad of the more physical of angelic beings? Have some of these, in past human history as recorded in the sixth chapter of Genesis, dared to intrude upon God's better creation here on this Earth from the sky into our atmosphere, the first heaven, Biblically speaking?

After Eden upon this Earth we have deteriorated so from this Utopia rather than learning from it and progressing. Rather we have so degenerated from such grand potentials and means to the limited and stunted frustrations in our known history.

Thus our Eden was of such renown that before our rebellion we already had, what we have since failed to achieve on our own over the great course of human history. Was it only gardens of awe-inspiring botany, or also a university of intellect, beauty and arts? Conceivably even part of a far more extensive series of architectural constructions as to be humanly unattainable, but for the Hand of God. If that was Eden, then we can now but glimpse our own vain efforts in science, engineering and creativities for the soul.

It was such a perfect environment that until our ancestors crossed a dangerous line, we humans had it all. Now we over the ages strive and blunder, grow and stumble only to make a partial progression; but as it is written: "a cloud that promises rain, but passes over" ... and over and over through this series of dispensations until God asserts so powerfully His ordained order. We all desire to return to Eden. Yes, whether we believe in God or claim not to, we all seem to want to return to whatever we imagine Eden to be. It is very personal and so intimate to each of us that we enjoy even trying in our individual ways to make and take time in our temporary lives to get back to that place ... so familiar, so wonderful, joyful and complete, but so elusive ... while even vowing to ourselves to never leave it again should we discover it. Interesting that we do yearn as Adam and Eve must have.

We are surely fallen children, but also experiencing the shadows of memories, such primal longings. We are so much like our original parents, and to our origins we still try to go as in the ancient of times, the Tower of Babel and other ziggurats demonstrated. Even today our technology is tending to go upward into the heavens, to our allegory of progress and beneficence. Truly we on our own accord are builders, building back to Eden, and in many ways literally back to the stars.

Our fleshly, cosmic dust with its drives and compulsions are masked under our social facades in the mode of society's standards. We pause, wait and try to restrain ourselves while underneath given enough freedom, even the most civilized and enlightened among us are but tyrants. I have never met a liberal or conservative who at some point, being as hypocritical as myself, is not tyrannical to a certain degree. It may be subtle with smooth overtones, yet still nudging and manipulating for our own gain. Without God we are all selfish to some degree.

It was in that past, wonderful Eden that we had all that we needed and desired ...everything provided, even healthy challenges.

Eden, its music and poetry, science and art, and its society were so pure and absolutely beautiful. But we just had to have our own way, rather than accept His provision, causing us to lose all of that for ages to come.

Human government has only shown how failed humans can govern themselves without God. Even with God, we tend to try to be independent, until we run into idolatry. This usually follows such a prosperity, where we more and more deny our need for God. This repetition has been the pattern in religion, academics, organizations of all kinds, sports, government and even marriage and families.

It was only after Eden's loss that we were vulnerable to those entities coming from the sky. In Genesis chapter six, these fallen angels had to come from somewhere. One interesting speculation was Alpha Centauri's Earth-like planet co-orbiting with our sun around this galactic home we call the Milky Way. Perhaps it was there amongst the Sun's other siblings that life first took hold.

Inherent Differentiation (IH) within the DNA/RNA soft/hardware triggered only by time and environment, affords progressive topological unfolding of biological geometry by God. In this second heaven (for ages before this played out) and in this time, our locale in this galaxy was to be the next stage for this proliferation of the miracle of life. As the somewhat older Alpha Centauri siblings became more benign to this newest of cradles from the local birth cloud of the sun, life had another seeding. Of the Galactic cradles and others beyond, it was under the twin suns here and the third red dwarf of Proxima Centauri in that other Earth-like sky, that God's intriguing story was to develop preparation. Here in time, beings of non-human characteristics, but with technology, would construct sensational things. These beings would be known later to us as the builders.

In further time, this Earth would develop and unfold natural life, as had many other worlds throughout the second heaven of space and time. Eventually, we have such a lovely world to inhabit and here our Eden would be. How long we in-dwelt Eden, how large the 'garden' was and how advanced its inhabitants is all unknown beyond the limited speculations of some. Perhaps we are now such a pale degeneration of what we were, that we would at present definitely be strangers in a strange land.

In this time undetermined, we had finally strayed enough to lose, at least for a time, our inheritance. Then as we struggled and tried to make a living by the sweat of our brow and by the pains of childbirth, we started to spread out upon the world stage of better known history. This record is still incomplete, but later on help would arrive that was not wholly benevolent.

Upon vast periods of time before within the Alpha Centauri System, beings became interested in other places to relocate. It had to be the right environment and surely some were hostile, but manageable. With their own developments of thought and measurement to survive and conquer their world, and even thrive therein, they had observed nearly 4.32 light years away, another quite lovely world. This measurement in light years was a variable and there were times of greater and closer distance which was caused by the Sun and Alpha Centauri stars orbiting not always in perfect synchronization. These variances allowed short periods of near constant closeness that lasted thousands of years.

First robotic probes had surveyed their local stars and would perceive the essential details. They were part of a far more ancient rebellion that resulted in their degradation from what was intended for the angelic host of God. These fallen ones were not the only ones in the vastness of space and time, but they were a local sample for the humans nearby to be one day aware of.

We had lost our protection and ease of life. We had lost so much in our own rebellion that we were far more vulnerable to what was around us than ever before. We had it all, and consequently lost it all. Now on our own we would try to rebuild our world. Some of us would listen to our Creator, but most would not. We would begin building things too, but that scheme would normally take centuries and patience. Just as today, we do not like to wait and are impulsive. We had gifts, but used them for our own selfish reasons. We just did not want to listen to God, even though it was only a short time ago, that we had had Eden. We found the way difficult and tiring.

The best of what humanity could possibly be was to strive for our utmost for His most beautiful of designs. This Eden that we had let slip through our fingers had what we all would spend centuries trying to seek on our own. It must have been much more than a garden. It may have had many gardens, maybe cities and regions. All that we have invented and dreamed had already been reality in Eden.

We were not supposed to die; we were designed to be immortal beings purposed to eternally glorify God. There was worship and companionship with God. There were surely other beings and entities too, animals and plants, mountains, rivers, valleys and seas. It was an idyllic, perfect world.

After the Fall, within ourselves pride in all of its subtleties in-dwelt as well as other nuances of the rebellious state. These behaviors would not go away for we now had to deal with our motivations and desires. Some of us were more patient and loving than others. Others were so dreadfully without conscience that the worst of tragedies was obvious. That was followed by armies and wars and all those miserable things that follow for sometimes generations.

Even technology must have been in Eden, as it does surface in the human species to use tools and improve one's surroundings. After the Fall, even the hunted had become the hunter.

Outside of Eden the climate must have been so variable that we were finding ourselves destitute for heat and shelter. The struggle for living, since God had sent us away, was enormous. We did not treasure what we had had, nor did we respect our God and His Holy ways that were meant to parent us. We were now found incomplete and still we defied Him.

While we struggled, unbeknownst to us, galactic entities must have orbited our Earth. Perhaps something would catch our eyes in the sky or conceivably we could hear thunder-like rumblings as well. But we were oblivious due to our intense motivation for survival.

At some point in time, what we thought was 'help' arrived. These new visitors of ours were so much more accessible than God, we had supposed; and we were so impressed with them. They were extremely intelligent and strong. They had even conquered flight and had come from above the clouds. The help we so desperately desired had come from the sky.

Chapter 4

The Builders that Came from the Sky

It had probably started quietly and over a long period of time, quite seductively and without alarm ... one here and one there, landing from out of the sky, isolated and dispersed amongst much large swaths of geography. Each remote society of early culture was quite detached from the others. Each group struggling to re-attain what was lost, now possibly only legend to some and to others a time before that never was. Denial is one way to not have to look back and consider the loss of Eden, their Creator-God and His Ways.

They finally had time and God's long-suffering patience to think that there would be little or no consequences for their defiance. While misguidedly considering God as a tyrant, they had already turned to their real oppressor, this Prince of the Air and Angel of Light, who was all ready to begin his next phase of interfering with God's Original Intentions for humankind.

We had only to obey. That is truly all our only Creator, Father, Benefactor and Saviour wanted. But humanly, 'pride always goeth before a fall', and what a fall we had perpetrated.

These visitors from above, with no obvious moral foundation had arrived. They were in the beginning so amiable, so eager to teach new 'inventions' and techniques. Instead of roving for great distances to find food amongst the edible plants and animals prolific in certain places, depending on seasonal weather variables, we were now beginning our progression into agriculture. Humankind began to use astronomy in order to track the seasons for planting and harvesting. We built cities, used mathematics and developed written language. We were growing in intelligence, self-enlightenment and becoming so progressive. But we continued to put God in the background as we were becoming more self-sufficient and self-important. 'Pride does goeth before a fall', and we had lots of pride to wallow in.

These unhuman like creatures were enormous and strangely unusual, yet they must have intrigued us. Some of the leaders, the more intelligent ones had long, arcing, rearward craniums. Some were towering in height and super-strong in size. Others appeared abnormally hideous. They fascinated us and we allowed them to more and more influence us. It was understood that they were involved in inter-species' genetic manipulations. In the twentieth and twentieth first centuries, such was called transgenic between differing species. God had warned about perpetuations of such abominations. But humans had for so long ignored any spiritual guidance that God allowed them to continue in their own deviant behavior. The visitors finally succeeded by trying to mix their genetics with our women to further the proliferation of their own 'race'. The novelty of their uniqueness over time had waned as their supposed beneficence was swept away by their demanding and controlling natures.

Stonework of grander dimensions resulted from this domination. Inherent human cultures of hunter-gatherers with villages of much simpler designs now found themselves hurtling into the science of astronomy. At first it was for their agriculture, but later used for the understanding of numbers and eventually trigonometry and geometry.

This was reminiscent of the almost forgotten past of Eden, for such things therein were common and familiar. Within the world-wide scale of Eden, with many gardens and sanctuaries, were also academic, literary, athletic, artistic and scientific pursuits. It was so profound that as the generations came and went, fact became unbelievable to many, simply embellished legend and later merely myth. Eden was forgotten and its truth and beauty minimized over time.

By the time the giants arrived, all things former had to be retaught and were simply degraded versions of the greater things that had been.

What God had originally intended was now corrupted and incomplete. Yet for humanity in rebellion, they appeared as 'new' things. Humanity demanded its godless entertainment, finding new idolatries to indulge in. Here it was handed to us, rather than taught carefully and within God's tender rules. At this point we gained a quick fix of knowledge and greedy power, while submitting ourselves under the tyranny of the dark angels from the sky.

It had taken years to traverse the distance from our sun to and from Alpha Centauri, and we were closer together then too. These grander cycles are so dim from the future to the past, as they are from past to future. Again, as the future reveals the past, the past reveals the future. On far larger scales of not only space, but also of time, are such things increasingly common in God's Universe.

It is a tenant of probability, the math and science of large numbers and certainties, that the rarer something is in a short space or time, then it certainty increases with a greater scale of space and time.

If one were in a desert and found a rose in some shaded, benign environment of rock, soil and water, it must be rare and protected. To the naive and untraveled, such a find would seem a wonderful and unusual discovery, one of a kind. But to the inhabitant always traveling and used to different environments, it would appear as merely another rose. But expand the space and time for that native, and the law of large numbers with the implied tendencies, and eventually our inhabitant would be at some place and time surrounded in an English garden with so many roses to his bewilderment.

It is the same with observing a bee, a mosquito or a bird. If any is noted that is a form of low resolution measurement, then there must be others, somewhere and some time. This works with patterns of size and complexity as well. There are significantly more of the smaller and less of the larger, for instance in regard to the number of sand granules and larger boulders, with the increasing spectrum of size to rarity proportional. With complexity, the simpler tends to in quantities overwhelm the rarer. Biology and technology follow these patterns.

So on a grander scale, that of our Galaxy, it should then be implied we have observed similarly on our local space and time. Predator versus prey, that which seems rare is more numerous, yet proportionally it remains that the smaller, simpler outnumbers the larger, more complex.

What has been going on in our Galaxy in its vast history in spaces and time? Our God is the maker of many worlds; He has in His vast creation within His space and time, powerfully extrapolated what we can only locally observe and measure.

It has been presumed that because we sense our limitations and consider ourselves only but a sampling, if on the Milky Way's galactic scale, then a very small part of the greater ecological system within our large home galaxy. It would have to be considered the same for other similar galaxies like Andromeda or Triangulum, the former, larger and the latter smaller, yet still spiral. This could perchance include the dwarf galaxies, like our neighbors, the large and small Magellanic Clouds. They might offer some surprising finds themselves. Let us try to just keep by similarity the law of large numbers, what is possible, based on the little that can be observed and measured.

Such are the extrapolations that are driving my intrigues. I can only 'see through a glass darkly', but I can get glimpses. For a desperate species returning to God from their defiance, they had become what many would agree, even those who claim not to believe in God, a degenerate form of their original design. In my elder time here and in the future to those before, because of technology being somewhat more resolute and presently attainable for me, I am getting a better view through such a clouded window.

Sumer may have been one of more such advanced civilizations as we think we discern it. In such short times, hunter-gatherers suddenly began establishing cities, commerce, writing and mathematics, intriguingly even astronomy. It is said that myth and legend have some factuality to seed from. Quite possibly over time, history of actual events had become embellished, yet within was the seed of truth. The Sumerians and later other civilizations appeared on the scene, as we look to the Fertile Crescent for advancements in academics. But what usually remains is another pattern -- only that which is the most viable endured, so implying what was the more delicate, from biology to anthropology and technology, slowly perished.

Everything leaves a trail, and as that trail fades tendencies of architecture and function are shadowed. It is non-intrusive time travel, when one is not interfering kinetically with formerly, inertially connected events during the present 'now'. Such is the wonder of these investigations, putting myself in their place as a desperate hunter-gatherer allowed me to empathize with the emotions of their dire straits. What one does after generations have passed and having heard about what was long ago so wonderful a place, would seem merely legend, to live and explore.

I had more detective work to do, as I wanted to better understand this history of our humankind. What so long ago in deep time, our ancient origins and then tragedies, the longings in each of us to return to our Creator and His Eden, was far more than poetic flight. It was sourced from deep within the human well of feelings, quite primal and difficult to resist. As such, a collective as well as individual primal memory within the human race, would allow a glimpse into the motivations in each of us.

To journey and seek sustenance, to survive and procreate ... then with agriculture to begin the base for a village, then in time a city with affluence, and with even more time, the finer endeavors of arts, sciences and commerce developed. Constructions improved from simple huts of wood to walls and buildings of stone. All had been of necessity to enrich survival and generate enough time for pleasures of intellect and art. Thus what we call 'civilization' was born, yet always our hearts deep within never becoming so civilized that armies and wars would cease.

Even the most ardent pacifist still has the capacity to lie, murder and steal. Perhaps it is in our history. Yet only a minority of any class of people actually commits crimes. In the most adverse of conditions, many are surprised to what depths the pillars of any philosophy will fall.

As Daniel Webster related, religion is a system of ideas. Thus all philosophies are a human system of ideas whose principles under testing, reveal their true values. For the Christian, then as offered, it is simply only the relationship each one of us has personally within the acceptance of Jesus Christ as God; not the reliance upon any human system of ideas. Humans without God have failed and have had to lie with much embellishments as a 'civilized' civilization.

All one has to do is test outside the paradigm of limits of assumptions previously untested by the local academics and socio-political established structures of our nation and those international similar hierarchies insulated from reality structures. In time, severe often subtle reactions ensured against a true assessment of the system. If the 'authorities' felt threatened, the tester found their work ignored or banned, or worse, their character vilified, even blacklisted and labeled treasonous. Civilization without God is mere illusion and dangerous confidence. Civilization is not far from tyranny and self-destruction, as evidenced in everyday events from one human being to a state, including all of the institutions within.

Civilization, is not as civilized, as we are led to believe. Test it, and one discovers truth.

Chapter 5

Non-Human Intelligence and Technology Precede Human

Along the long savannah trek the small family had come, near deserts and jungles, ever trying to stay in the savannah's safer places. Here one could observe for a distance, quickly build shelter, hunt and gather food, from meat to fruits and vegetables.

Far away one could observe fires, herds and weather, perhaps finding others with whom to associate or trade. Each family numbered many, even joining extended families were the norm. The tribe then became the next foundation of society after the family.

Slowly, very carefully by trial and error, each tribe would share within and sometimes with others who would ally with them to best survive. There were occasions when a group of tribes would stand together against another group, thus wars and their resulting triumphs or tragedies would bind them together even more.

It was at night at the communal fires that the stories of long ago would be retold and embellished, so much as generation begot generation, some of the stories would enthrall or frighten. Many of the tales were so magnificent that one would need to dream upon them in the quieter moments. When other tribal groups would gather with them, then more stories would be shared. What was the long ago time like, where all was so simple?

But lately, there were stories about their own time, not long ago, that some tribes shared. These were about 'teachers' and 'birds of fire'. Some said other birds made slight noises and even some had lifted some people to the sky!

There were stories of medical help for injured ones, some type of interventions upon warring tribes with powerful medicine techniques. Some of the women were bearing 'new children' and some of these were often very ugly and strange.

These tales abounded and seemed to be about a place far away that required a long journey. The adults would be so awed by such tales that after the great meeting many had to enforce some sobriety on all of them. The savannah was a dangerous place to be careless in your watch and necessitated a sharp edge to survive ... mentally and physically.

But in the quieter times when the tribe was more laidback, those stories could intoxicate one's thoughts. For the children it was not a problem, but for the adults, especially those in leadership, it was slowly affecting their decisions.

As the months passed and seasons changed, and when hardship would require endurance, the tribal elders considered more and more to journey slowly towards where the 'teachers' were. It was getting closer as their migrations headed towards that horizon.

One particular morning sounds from the sky lingered as thunder, but not going away as fast. The sky was sunny and blue with clouds, but no storms in sight and the warmth of the day was increasing.

Stars had always shown at night, and yet recently in the daytime sky were some stars, usually three that moved above the clouds and did not twinkle. They were 'day stars' as some thought to call them as they moved together and so easily caught the eye. Later on in the morning, more day stars were seen, and then in the afternoon came an increasing thunder.

At first it was difficult to locate the disturbance. At one time it would be toward the east, then the west; sometimes it was encompassing with no discernible direction. But it was growing louder, with a rumbling sensation also in the ground becoming more and more noticeable. The vibrations rumbled below, the thunder was all around and suddenly the wind was whirling dust and sand into a cloudy torrent of frenzy.

Fear suddenly gripped everyone. A humming, whirring sound had become so predominate that everyone in the tribal group hugged together, forming a cluster for safety from that which they did not understand. Suddenly, before them appeared upon the ground just meters away a metallic, ovoid shaped object. Was it a strange flying creature that was hunting them? Then just as suddenly, all the wind and noises died down to a very calm, quiet that surrounded them.

Everyone was so horrified by all this that even the hot afternoon Sun upon the savannah went unnoticed. Slowly, the side of the ovoid opened and something was sitting in its doorway. Down the steps came some strange, human-like creatures that were very bulky and small headed. So pale they were and not attractive, quite ugly actually. They stood on each side of the steps and slowly a very frail, old form was let down above the steps. It never walked, but seemed to be carried. It was even uglier as to cause one to look away. Its head was large and elongated to a curve rearward. All of these creatures had no lips, but teeth showed through even though the mouths appeared closed.

There they all remained, the tribal group staring and silent and the strange creatures also noiseless, returning the stares with very piercing eyes directed into the eyes of the others. The tribal people just as suddenly averted their eyes and could not continue to look upon these creatures, so ugly as to cause fear. Then from out of the surrounding air came a voice that was from the old form near the ovoid object.

"My children, do not be afraid. We are here to teach and help you. You must be tired from your travels. Are you hungry? Are you thirsty? Let us know. Come forward any of you when you desire, for we have time and are patient. Let us face each other and share and work together. Life is hard and can be easier", said the frail older one with a slight smile-like expression. As this leader spoke, his eyes continued to probe and penetrate those of the tribal ones. Then lifting a frail skinny hand the old one beckoned with a gentleness, slowly and repeatedly motioning for anyone to come forward. Behind the voice was a measured, monotonous breathing. From time to time the ones along the steps could be seen drooling from their closed but leaking mouths. Even the blinking of their eyes was slow and infrequent, as the old frail one would gaze from side to side just as slowly.

The afternoon seemed to go on forever as the still air and the hot Sun beat down, yet the closer to the ovoid, the more comfortable it seemed. Soon the ones along the steps, like guards, unfurled a tent-like sail and protected the tribal ones from the heat of the Sun. No one spoke from the tribal people as a silence ensued with just the continued, tense scrutiny of one group to the other. One side trembling, and the other so confident and composed as to not have a worry in the world.

Softly, a quiet flute-like song whispered in the air and seemed to help calm the ones who had traveled for generations on the savannah. It was so refreshing to be offered, without any effort on their part, fruits, vegetables, meats and cold water. Something in the dark liquid presented was so soothing, but a bit intoxicating if one had more than one cup. The afternoon passed until dusk was approaching and more of the people of the tribe felt accepted and at ease. It was a very welcome, cool oasis that was offered by these strange creatures, proposing to teach and help these struggling humans who have had to provide for themselves all along. Some were sick and lame. They had a very hard existence. Here was such sudden respite that it was enchanting to say the least, for unbeknownst to them, that was the intent. These must be the 'teachers' that the tribal people had heard of. They were quite overwhelming yet so beneficent.

Slowly, more and more of the tribal members, one by one or in a small group, ventured to taste and enjoy the gifts offered for comfort from the 'teachers'.

It did not take long for them to realize that the old, frail one was the 'teacher' and the others were 'guards'. They would learn also in time, there were 'workers'. These were all 'builders' that had travelled very far in the sky. They had much to teach and had so many strange things to share. To learn how to plant and gather food, including raising animals, was the essence of agriculture. It would require knowing the seasons of the year and reading the sky. To store food would require building, and to trade for other foods and materials to continue such a regimen, would require a sense of commerce. Mathematics and writing would be needed as well. All such had been lost since the Fall from Eden. Now humankind was on its own as it had wanted.

Adolescence without restraint is dangerous. Satan had sold a bill of goods that cost humankind its fully functional state. Now degenerating by each generation, God's Utopia was gone and now another was usurping the fertile species that was created just for God. Many there were that also followed this far wider road of least resistance. Only a few would attempt to seek out God's narrow way back. As in history it is a few that are right and the majority proven wrong. The 'blind were leading the blind', and both had fallen into the pit.

Non-human intelligence and technology existed not only amongst the stars and other planets of Alpha Centauri System, but now on Earth as well. Through intoxication and seduction these darker angels, these men from the sky, these giants, would slowly subdue the 'lost' of humanity. The worst was yet to come.

As a few days passed, the humans of the tribe were welcomed to join the teacher and to journey with them back to the city so full of delights and wonders. Many wanted to go, and yet there were a few of them that wanted to return to their former way of life. The old teacher seemed to be so considerate and allowed those who did not want to go remain, and even more so encouraged those who chose to follow. Those remaining were younger ones and the sick and elderly. The rest were so warmly invited to go where the teachers and amazing things were located somewhere beyond the sky.

As those who chose to stay behind on the savannah watched the ovoid begin to stir, whirr and lift dust to swirls upon swirls. As its humming song resonated all around, it rose into the sky as a day-like star, glowing and then moving on towards the horizon.

The now smaller group of hunter-gatherers of the tribe looked to one another to try to reorient themselves to life as they better knew it. Later the night would come and they needed to settle and camp, to survive and go on … alone. Off to the distant horizon they could see the sky slightly lighted. That is where the others were. They wondered about all these things, but not able to explain or understand any of it. Some of them would scratch into the rocks, clay or wood around them images of the unfamiliar, while others were considering going the next time.

The ovoid was strangely comfortable and fantastic to travel in. All gathered together with enough sustenance that everyone's hunger and thirst were satisfied. They were offered a plate of sweet foods to try and with many now becoming accustomed to this new way, they seemed to relax some and actually enjoyed these new things. Some wanted more of the sweet treats that were so tempting. It was as if one could not resist to sample some more as time went on.

Later, they realized they felt no movement and were suddenly aware that they had landed within a great city. The door seemed to silently open revealing unfamiliar stonework, buildings, columns and streets never before seen or named by their kind. The smells therein and the sounds of music were overwhelming to them as they were invited to slowly stroll out upon this new vista.

The Sun had set and they saw by lights and technology in the immediate distance, this city they had heard about. All of this was so intriguing to them. The younger ones especially welcomed it all. Never had anything like this been seen or experienced by any of them or by any they had known. Unhurriedly they gathered around to prepare for sleep, if it could be possible. Everything was comfortable and beyond adequate and one by one they drifted off to sleep.

Chapter 6

Ancient Visits upon an Earlier Earth

The co-rotation of our sun and the Alpha Centauri stars has been a very long continuation of the same cosmic dance. As the small open cluster produced these same stars, their separate progression was one leading the other. Alpha Centauri's two main stars had always been matured, about one billion years ahead of our sun. In this cluster were some large stars that in time would be supernovae, imbedding heavy iron nuclei into the early materials of other stars near them, including Alpha Centauri, our sun and their planets. Proxima Centauri's very large orbit had at times brought it much closer than the inner two Alpha Centauri stars, which affected and disturbed the comets of the past, and will again in the future. This dance would have them leading and then lagging the Sun, and coming closer and then farther away, back and forth through millions of years.

These same millions of years would also see the maturing Sun and its planets, including the Earth. Planet migrations of the giants, Jupiter, Saturn, Uranus and Neptune with collisions and Neptune very inclined to the solar orbital plane. Massive periods of inner solar system bombardments of asteroids and comets, including giant comets would be of the norm periodically in the solar system history of the Sun. There were similar events taking place in the Alpha Centauri inner stars, and of a lesser nature for Proxima Centauri, yet before one billion years earlier.

Continents formed from early island chains in Earth's ancient oceans and ice ages coming and going. Atmospheric changes occurred over time with no oxygen, then some in varying amounts so great that larger animals and insects could exist.

Florida, once nearly recognizable to those of us in this present time of the twenty-first century AD, would over time vary; as an archipelago of islands forming linearly from the northwest to the southeast, to other times a land mass around three hundred miles wide with continental shelves and grasslands with herds of now extinct life.

Occasionally as the solar system of the Sun grew more stable, an errant asteroid or comet would crash upon the Earth causing massive climate changes. Even more rare, as our sun and the Alpha Centauri stars coursed together in their Galactic orbit of about 250 million years, was a supernovae that would inundate the worlds orbiting these stars, affecting life there on.

While Earth was the more primitive and younger, the Alpha Centauri stars had the approximate one billion year lead in development. There in God's grand designs, his mysteries and His sovereignty, other life through Inherent Differentiation (ID) would attempt to survive and overcome its environment. This ID was now what scientists would consider to theorize that life is already designed and inherent within its DNA and RNA with all it can be. Already within the DNA/RNA hardware/software programmed by the Creator, it requires only time and environment to trigger the unfolding of the topological bio-geometries proper for such adaptations. This is true for here on Earth or any other planets among the stars, obviously including the Alpha Centauri stars.

Already within the seed of a rose is the rose's future; the seed is the past for the rose of the present. And thus with all life, the temporal explains much of the mystery that seems to have perplexed sages. Here and now, the mechanical theories and randomness without design, so-called evolution thus fails to explain. For they had forgotten that time, as a fourth dimension answered all that they could only theorize while denying God. Now considering time with a block perspective based on Albert Einstein's interpretation, it would allow determinacy and no place for randomness. For as in any physical process there is a logical progression, thus randomness is just the lack of resolution of measurement. Only a determinant-based universe can function. Time is the disregarded part of the story, for function that can be of evidence repeated and not fallacious, fantasy is unprovable.

Cycles within cycles, as Ezekiel's wheels within wheels, but for far longer periods of time and greater distances in space, are that which appear so rare, nonetheless far more certain. Probability Theory and the trends it implies are so fascinating pertaining to the things that were, as for the things that will be. Thus what appears as a presumed, immeasurable 'random' is more measurable with less refinement of resolution, for it begins to repeat with the far greater dimensions … the three dimensions of space and one of time.

To have visited our familiar Earth, thirty-three million years ago during the Eocene or sixty-five million years ago in the Triassic/Cretaceous or how about 250 million years ago in the Permian … we would have found a considerably more alien planet than what we are used to. On into the future of such great time periods and again we are viewing a more unfamiliar and less comfortable planet. With ever changing night skies and varying constellations of the different stars the more farther away, and nearby for some stars that are swiftly passing through our area of Galactic space at those times. Past and future, and especially far past and far future, to our present is strange and would be to us from our present here and now, as if we were visiting another Earth around another sun-like star. Over the far distances of time and space, it would be almost irrelevant to what was past or what would be future. With the increased repetitions of the rare, and of course the more familiar of cycles, it becomes more like the quantum view of neither past nor future … very little of reference left for how we reckon time and place.

Eternity seems the norm in these more extreme time and space dimensions of measurement. Eternity is not so far away in our normal macrocosmic world of everyday, but just far more elusive to observe and measure. Break an egg, or break a piece of glass and that action should be impossible to turn back. So what has happened is forever marked. Everything leaves a trail of forensic evidence to indicate the arrow of time. But that resolution is more difficult to obtain with the greater cosmic scale distances.

Similarly within the quantum world, where things are so miniscule and happening so fast, the forward and reverse time seems to blur into insignificance as the microcosmic scale of things bends our familiarity. Here whether 100 million years in the past or into the future, or in our kitchen lab to a place 100 million light years away, all suddenly seems the same. Which way is time going when we have no familiar reference points temporally?

The same for space, for if one just imagines the cosmic as an equivalence in perspective to the microcosmic, then just speed up the rate of time for the cosmic and we then have a better understanding of this comparison; likewise for the slowing down of rate of time for the microcosmic to simulate the macrocosmic. Do we have perhaps a better view of what appears so different, as more the same?

Time is so important, as we have discovered in understanding that which is so often presumed to be chaotic, 'counterintuitive', noisy, random and turbulent. Thus it appears beyond the available resolution of measure of our personal here and now.

With this basis, how often has our Earth, in the past and in the times to come, been or will be visited by such fallen angels? Is the universe rather bait, more predator and prey as is reflected in the natural world? Has God made man and woman, His humankind in His image, so much in need of His guidance and protection?

Perhaps He has ... from what we can determine from this single probe of the Alpha Centauri System, and extrapolating upon all that we have most recently learned. With these most recent clarifications of the scientific records we can better filter and readdress many of the past, incomplete hypothesis from the last five hundred years previously thought to be so complete by unscientific presumptions

What had formerly so divided religion and science, now seemed to unify. The future was revealing the past, as the past reveals the future. Given the circumvention of the convenient presumption of enlightened yet unscientific dogmas, then humans, left to their own will, overwhelmed the more patient and scientific. Without or oft times because of the avoidance of ground testing, those feeling threatened by certified repeatable evidence quell the truer explorations, thus becoming academic tyrants. As Shakespeare well said, "Sic sempre tyrannis", also to those who for their personal and selfish reasons, hinder true scientific pursuit.

Contemplating further back into times of prehistory, thousands to even hundreds of millions of years ago, what else might have visited the Earth, and from where? During inter-glacial periods, Earth's lush gardens with its spectral bio-signature must have beckoned for light years into space. Following the extrapolation of what we have discovered about the now desert areas of Alpha Centauri's B, its orange sun's Earth-like planet, perhaps others of the past had stopped and visited our solar system and its environs.

If there were any remains of such mechanisms and non-human intelligences (more fallen angels or perhaps of a more benevolent kind), then those things would now be deep in some of the sub-ducted continental plates or even in the upper mantle. Some items would be so corrupted by time and the elements that only a small indication of dust of an unusual nature would remain. What if they had built something? Well, as the Oko 'reactor' in Africa possibly indicates, they might have had ancient nuclear power. Of course that may be just a rare conglomeration of uranium by water and other normal geological processes. But it is a very interesting artifact indeed.

We have on our Earth other fascinating finds of construction that are older than even the pyramids. Those massive structures familiarly come to mind, and are well to be considered, but also the Sphinx, as it seems much older than the renowned pyramids. Across our globe there seems to be many more unusual and intriguing artifacts of very long times ago such as the Bimini road in the Bahamas and in Mexico the Aztec ruins.

These massive examples of architecture and others share unusually large stone works with heavy construction. Stonehenge in the British Isles and some those aforementioned are often found to relate to astronomical qualities. Here we also include the Native American sites of the southwest and mid-west of the North American continent. Such extrapolations must be based on as bizarre evidence as that of the fantastic conclusions being offered, similar to the evolutionists' way of trying to establish interesting theory as fact by implication, but without repeatable evidence.

Have these Alpha Centauri builders or fallen angels been here before? Or have there been other 'angels', fallen or benevolent here previously, even before the Alpha Centauries?

If we would take the proverbial sample of our local space over great periods of time, interpreted from our limited time and the best of repeatable evidence so far gathered and reflected intelligently upon it, what implications might we consider and what trends might we apply? What would this sample say of our Galactic ecosystem and its eco-history on God's scale of things? This dares not to extend to other galaxies like our own, but is implied.

It seems more than enough to contemplate this grand perspective seriously. It seems in this sobriety of academics, that God had protected us and had created us as very special to Him. We had rebelled and found ourselves thus separated enough from Him as to be vulnerable to what else was in His great Creation on the Earth and amongst the stars.

We were so easily infatuated by the builders, the fallen angels and the teachers, that we then as now enjoy being taken care of, having our pride stroked and taking the easy approach to comfort and power. We replace God in our fallen nature with other gods, including ourselves. We prefer idols and other non-convicting entities to our Creator-God, to the point that even these 'men from the sky' could so easily entice us. In the latter twentieth and early twenty-first centuries, we had thus allowed unethical businesses and governments to use advertising in all its latest technologies to so influence us. We are supposedly more educated as consumers than to consider things thoroughly, that we even after all these centuries are still seduced, entrapped and misused. It is said that there is no honor amongst thieves, well the same for predators. And in a world outside of Eden, the predator versus prey would prevail.

In Eden we were as humans at our highest pinnacle of development. Ever since the Fall we have attempted, on our own or under the influence of others with no moral convictions, to build our own versions of Eden.

The long term of ages and cycles amongst the stars, and this sampling of such space and time upon the Earth, revealed far more of our past. Here in this future time with our advanced resolution of technology, observation and measurement to those ancient, prehistoric times, we could better understand the tragedy of earlier humanity.

A probe that now was transmitting from the nearest of sun-like stars, around 4.2 light years away, was revealing ample drama of such a personal nature. I, as each of us should acknowledge, am a descendant of those before. We are the ancestor of those to come. It is written that we have such a 'great cloud of witnesses'. Now I see that the witnesses are not the ones nearby, but may be observing from a more secret advantage of place and time. Far into future time and distance, with greater resolution of being able to discern events and objects of long ago, it can be presumed that others have also observed us.

Often in history, events of great importance are considerably enhanced from more recent discoveries in archives or in areas of past conflicts of thought. For instance, Custer's Last Stand has now been more lucidly interpreted, revealing by modern forensics as more Sitting Bull's Last Stand. More than one hundred years later, evidence of what really happened there was made clearer by the clusters of bullets and remains of the battle. What was handed down by the Grant Administration, blindly following others in authority was a very biased and embellished story, later denied by evidence to the contrary.

Similarly with former data-based accounts of ancient Rome, particularly Pompei, prehistoric Egypt and others have become far easier to comprehend with better clarity. Thus the same can be said of this information from the first known human, interstellar probe. Yes, first known, for I wonder if we had not gotten to the stars long before, and this is another attempt to achieve great things without God or by keeping Him at a convenient limit. Alpha Centauri has beckoned for decades upon the human imagination of what it is like around another sun-like star.

I had to get some rest, it was all so fantastic and yet tiring, emotionally exciting and draining after a while. The repose felt so good to contemplate, but how would I slow down or stop my wondering. As old questions are answered, I gathered new ones; queries satisfied and more to ponder. How much more do we need to learn? What will we learn that might surprise or even overwhelm us?

Suddenly, it came to me that not only we in the space sciences are so involved with the discoveries of the greater universe around us, but so are other experts. Archaeologists at first glance might not appear to be studying our place and time in space. Their 'space probes' and telescopes rather are brushes and trowels, dust and stone, exposing layers upon layers being patiently and carefully studied; bits of parchment, pieces of pottery, leftover food, so hidden and dormant for thousands of years.

Going father back we enter into the realm of the anthropologists, who now have a more complete history to reveal the slights of information of human origins. They too are digging all so delicately.

Then we get to the paleontologists, those who go the longest way back into life as we know it. How little does remain of anything the older it is. Entropy, the forward direction of time, does leave its wake of consequences. It allows us to measure things quite well.

Then it is the geologists, who try on a far larger stage to integrate even more powerful and larger dramas. Now we are back to the cosmos, for that is the theater in which all of this has occurred over eons of time and deep space. Beyond this is cosmology and then mathematics, particularly, the spectrum of geometries; then philosophy and religion. Religions there are many, but only one in which God is the Beginning and the End, the Author and Creator as well as transcendent, yet personal.

How many times Earth might have been visited, not only by the giants, dark angels, builders, teachers or others, remains specifically unknown. But if we consider the implied trends in mathematics, probability may suggest that it would be more than once, but less than many. Such is qualifying that which has not enough information to quantify so firmly.

The next morning I awoke and took my time preparing for my day. A little boy was fishing at the lake outside the window of the home my wife and I stay in here in Florida. I watched as he would cast his line one place and wait, and then another and wait for a bite. The float, only a little plastic air filled ball, was his signal along with the tugging on his line. Such is also SETI, the 'search for extraterrestrial life'. As the boy 'listens' or better 'feels' for a fish, we also listen for radio waves with SETI. And with OSETI, optical search for extraterrestrial life, we look for laser signals of approximately *1nS*.

Hunting and fishing lend themselves to a useful analogy here where one knows and sees their prey, but when our prey is invisible, we must bait and wait. But if the extraterrestrials already know us, without our prehistoric knowledge, then they are more like hunters. It was a predator and prey game, benevolent or malevolent. Seduction, deception and even baiting could have, and may very well have, been used.

If such creatures, all created by God, specific to their own estate, were permitted to interfere with another, then this may have been what happened, in Genesis chapter six.

Chapter 7

Limits of Humanity

I had taken some time off and gone to the university nearby to relax on the grounds, just letting go and affixing my thoughts upon just the everyday life.

"Good morning, Professor, and how might you be this fine day?" asked my biologist colleague who had taught here for a long time and was planning to retire soon.

"Ah, well, taking a break from the data analysis of the first interstellar probe at Alpha Centauri," I responded knowingly stirring a comment beyond the mundane.

"Hmm, quite a reply, Professor … 'tis intriguing!" returned the biologist, with a smile and twinkle of eye.

We had quite the conversation about what used to be called convergent evolution and had been replaced by Inherent Differentiation (IH). IH now solely based on the topological mathematics unfolding of imbedded geometries in time and space, not only of matter in its many simple to complex configurations, but of life as well. So life, based on this mathematics, not on non-repeatable implications, infers convergent, topological unfolding of similar biological forms for similar mechanics required by environment. Thus with just two triggers, time and environment, the inherent programming's algorithms in the DNA/RNA allow for what exists in all possible conditions of time and space, as well as the benign of environment.

Consider studies over the last one hundred years or so of insects, dogs, humans, flowers, trees and single to multiple celled animals and plants … without any more than mathematics. From the perspective of our Galactic ecosystem over twelve billion years, the carbon being finally made in the stars, life is mathematically-based as inanimate matter. In non-living matter, from atoms to molecules in all conditions, convergence shows up in the patterns. From cold to hot or near to far, the cream on top of hot coffee swirling, to a galaxy spinning … all infer being based on repeatable mathematical evidence in the laws of chemistry and physics.

This then can be implied for life, for there is design in the patterns and numbers. From the later nineteenth to early twenty-first centuries, ignoring mathematics so as to bolster a 'randomness' to life was the norm. We now have a mathematical base to Inherent Differentiation (IH). That DNA/RNA was programmed as long ago as twelve billion years for our Milky Way, perhaps longer for the older galaxies, by only two triggers -- time and environment -- for the adaption of life in all its forms.

We have a Creator that is beyond us in time and space and has enabled the physical part of life to proliferate amongst the stars and perhaps amongst the galaxies. His geometries are difficult to perceive, but the most essential of foundations for life we see in the universe, His natural world of so many worlds. Much as the coffee's cream spins so obviously observed, yet it could not exist except for the rest of the deep coffee's heat and motion noticed ever so more subtly.

This was the revolution of Hypergeometric Mechanics of the early twenty-first century. The universe we so easily observe and measure has often been the cause of our inability to perceive and measure that which is cloaked, yet still there. That which is so hidden requires far more sensitive resolution, being of far more importance than that which is obviously yet crudely perceived.

Our conversation continued with some other friends of ours at the university, one an astrophysicist who was quite the skeptic too.

"So many from the creationist and evolutionist side were so busy coercing their point of view upon the other, that repeatable mathematical and scientific evidence was not allowed. Each side cherry-picked at facts in order to bolster their incomplete arguments," he offered.

"Look at the incoherence between the quantum and relativistic schools for so long. It was the same thing, where both had incomplete theories and only repeatable evidence for each side. Thus neither side ever overlapped in understanding. Part of the spectrum of repeatable evidence never considered," he continued.

"Why, it was in Riemann's perspective upon the prime numbers, and also taking seriously the Imaginary Numbers, where the great completions were done. How easily to just dismiss and accept definitions for the square root of -1 and just call things, 'zero, infinity', and 'chaos, random and turbulence', so as to not admit that time's inability to calculate with a high enough resolution those things we could only glimpse," I added.

"Of course", continued the astrophysicist, "with academic egos, politics and funding at stake, it was easy to not be challenged. But then came the studies from the Hypergeometric Mechanics that you had mentioned. Yes, then with the Temporal Diffraction Grating, the Hyperplane and the Hypersphere, we finally could begin to get enough of a glimpse to analyze that which had been fostered only by incomplete arguments."

This probe to Alpha Centauri was the first, and quite lauded when it arrived after over forty years in transit. Currently our spacecraft were expelling the Event Horizon of what we now call our superuniverse. We can better explain the red shift as the severe curvature causing the galaxies to appear receding at near light speeds, to only realize that it is our observational universe curving over the Event Horizon of our superuniverse.

Of course we find this all around us in a three dimensional space; while if we add time as the fourth dimension and Inertial Geometry, which used to be called 'gravity' as the fifth dimension, then we have a 5-D sphere. Mathematically this reduces our spherical universe to a flat 2-D disc on this superuniverse sphere. There are other 'local' universes upon this greater sphere, also reduced to 2-D flat discs and having very bent curvatures at their edges.

This surface of the superuniverse, rotating and expanding, is the speed of light and is the Event Horizon of our superuniverse. All event horizons are no more than local spots of the superuniverse's Event Horizon. When any matter crosses through the more frequent microscopic to the rarer cosmic, and very rarely the macroscopic, energy is traded ... and far more than that too.

It made sense because of the great energy in the nucleus in atomic power, also in the quantum world's mysterious qualities and for the superluminal that cannot be explained in space, after the Lorentzian effects in relativistic velocities occur.

I had pondered all this after our discussion that day and of such stimulating conversations of scientific thought. But there was a slight difference in these times of more complete insights. For we shall never attain absolute understanding.

There was a humility to those of us in the mathematical and scientific professions. After so many decades of humankind telling itself of all that we could or would do, we rather have realized our limitations.

The biologist had so many years of research upon the nerves, ganglia and brains of insects, dogs and humans. He spoke of a quantum-like ability of groups of insects, birds and herd animals seemingly able to all at the same time act and react in unison. Fish do it and even humans can do it.

For example, the brains of an ant or house fly are so small yet so capable of amazing abilities in very fast time. It has been considered that they do not need a Saviour, for they and other creatures already hear the voice of God and obey. They have not known disobedience to their Creator. We humans are the only ones to choose that. Some non-human creatures are made for good and some for evil. God made the dark angels, the devils and the giants as well as the moral beings.

Other mysteries, principally the human ability to have a foreshadowing of a future event or to feel familiar to a place and time that only their direct ancestor had visited. And for others, to just wait upon Him that had created all of this. How could things be where coincidence no longer was in the numbers, but rather certainty? The great limits of humanity were only transcendent when in synch with the One who is transcendent, a conclusion many of us accepted.

When such high resolutions of repeatable measurements became something precipitously to reckon with, then others were won over. With such a probe sending us from great distances of time and space, data of such implications vying to clarify our pasts, we had much to reconsider.

Are brains also quantum machines too? Even a camera or a sound recorder is a quantum machine if we allow that their information is available from a past 'now' to a future now and 're-visitable'. Then they are non-intrusive time or time-predominate quantum machines that are not kinetically interfering between the two times of their focus, the origin of the incident of what was recorded, and the re-play of that incident.

Again the human is such a limited creature of time and space; but with remarkable gifts from the Creator we are able to glimpse with new technology more of the generations in the future to the generations past.

Even with Hypergeometric Mechanics, the corollary of 'two different times at the same place' and of an atom passing the time of one second per second is 'two different places at the same time'. Then likewise, when an atom is on both sides of a barrier in a tunnel diode during the same second, all of this is subject to limitations. For only God can be everywhere all the time. This is the quantum world of the atom. So now replace the atom with a macroscopic human in a macroscopic Hypersphere, moving from the macroscopic de-coherence to quantum coherence, and you have the same as the above example. So we are very limited, and that is so very wonderful, for it allows so much more to be discovered rather than to be assumed.

Some days had passed for I had gone on a short but much needed respite. Upon my return to the data building, I could then with the help of others, return to a better clarity of substance. Some additional information was awaiting that prompted me to call a meeting with another colleague in the paleontology area of studies.

"Hello, hello!" along with a knock at my door, came the paleontologist.

"Please come in, sir, please come right on in," I responded with my out stretched right hand offering for his.

"Yes, yes, thank you, thank you," he said as he shook my hand quite vigorously.

"Well, I have looked over your conclusions, and based upon the data we are in a Second Miocene Epoch it seems."

"Yes, yes," the professor continued, "there has been increasing warmer weather, pattern changes and occasionally cold snaps over the decades of the latter twentieth and early twenty-first centuries. This is a pattern from around twenty-five million years ago until about five million, in which we went from very warm to much cooler weather. It really has been so beneficial for us frail humankind, by God's grace to not only exist, but to thrive. Normally over greater expanses of time beyond the historical especially, the Earth has not been so kind to us. We humans are quite fragile."

"Yes, my friend, it seems according to your discourse here that there is a possible fifty million year cycle involved, and so around five million years ago, the warmth gave way to a cooler temperate climate, and that is what the limits of human history has only experienced. Am I correct?"

"Yes, oh yes, quite, quite," the paleontologist replied with a smile. "We have always thought Earth to be a very habitable planet, and have history as the baseline of climate. But such is not so. I have looked at the conditions of our solar system on its orbit around the Milky Way and so it seems that much of our early history has been very affected by our Galactic as well as our solar system environment."

"Then this Second Miocene Epoch could last another twenty-five million years of warming, then a cooling period for the same?" I awaited his answer, with my hand thoughtfully tapping upon my chin.

"Well, we have to remember from our exoplanet studies, that to an analogue of our Earth, even with very similar animals, plant life and other conditions, the level of oxygen and the amount of continental crust is very important too," was his answer.

"Oh, I agree. The planets we oft try to call 'Earth-like' since the latter twentieth century, have been quite different in either mass or diameter. And with differing atmospheric pressures as well, the amount of water has been of importance too," I added.

"This Earth we have known has a history and it is a continuum as all things are. With our continents in almost similar positions, conditions may be repeatable for another Second Miocene Epoch. Eliminate humankind's agriculture and industry, even all of humankind itself, and then we have the volcanic episodes of major eruptions or an asteroid or comet impact of a rare but major sort. These also would be blips you have on this continuum, but the trend of this would, I conjecture, continue. It would appear as another similar Miocene," he instructed.

We chatted for some time, and when I offered tea or wine, he selected a merlot. After an hour or so he left and I looked up into the night sky, then slowly back down upon our Earth. How we have learned so much, perhaps only glimpsing trends and not so much the details. How our limits, no matter how we ignore them, are as real and decisive as our potentials.

For the time I had intended to relax, it had been quite the influx of food for thought ... different perspectives upon which I would ponder.

Human history, after the disappearance of the giants or angels from the sky, had over the course of time seemed more myth. There had been no testing by us of such things since those times. Without ground truth, any theory can survive until either confirmed by repeatable and verifiable evidence, or finally becomes invalidated.

As we are still in the Cenozoic, no more in the Tertiary, and quite a bit past the Miocene, I enjoyed the extrapolation of imagining that past was future. The thought of considering we might be entering a Second Miocene Epoch in the late twenty-first century, though seemingly just fancy, also made me wonder if, perhaps the giants or dark angels had non-human industrialization in a far prior time. Could it be that in some future time, a post-human time could be again, but of a non-human technology? Could it be more of a certainty, that as we are able to observe and measure the limits of present time and space, humanity's perspective on the universe becomes clearer?

Prior to God creating humans in His image, were there non-human intelligences whose advanced technology allowed them to traverse the stars, at least those near to them?

Sometimes, just asking questions such as these are as unnerving to opposing viewpoints as just stating possible realities. I have often said, all one has to do is ask a question be it right or wrong, and enough of a reaction can be excited in another of an opposing view. Regardless they will get so insulted, no matter how intellectual or academic they claim to be. All we can propose and substantiate is repeatable, verifiable evidence to support our differing perspectives. And to ask questions with the caveat of only such scientifically, reputable evidence, will allow a disclaimer of 'objectivity' to embarrass themselves into their religion, which according to Daniel Webster is 'any' system of ideas. So a religion defined as only having a deity is as valid as one that denies a deity. Religion can be thus only one's own system of ideas or beliefs. Many vie to convince another of their point of view. But without sufficient support of debate, they fall short with only meaningless rhetoric between the views, both lacking enough to claim the win.

I have better things to explore, than to waste my time with individuals who only argue to further their cause, yet cannot support it scientifically. Oft times they merely quote others with similar viewpoints and no objective research. Then they are so stubborn with denial, that when it comes time to ground proof their 'facts', they cowardly run and hide only to nurture their limited view and argue again. If they believe in evolution, then their religion of evolution is trying to bring them to extinction; for they and their ideas need to be protected from anything that is considered adversarial. So like any truly, unprovable concept, it in time and environment is not viable. Such need that 'teddy bear' of falsehood to never be tested. For then their hidden agenda, usually against God is able to continue. For me and many of our time in this latter twenty-first century, we have the findings and conclusions those before could only speculate upon.

The added perspectives certainly were fascinating. The Galactic ecosystem of which we were aware presently, especially in relation to human anthropology, archaeology and history, had been now placed on a far greater theater of time and space. God's perspective must also be on such grander scales, as well as smaller. No wonder we humans are so limited and replace what we cannot understand with our egotistical re-definitions in our studies and opinions. In mathematics and sciences as alluded to previously, the labeling of concepts that we cannot observe and measure we use 'zero, infinity, chaos, randomness and turbulence'. I reassert the same for the numbers labeled and only considered 'imaginary'. Such great vistas have been stifled by those in academic tyrannies to keep their views succinct while debunking any others of greater, innovative things. Not even allowing a hearing to even listen. Oh well, it seems so true as is said by some, that 'we should have been to the stars 1000 years ago'. But in our time of defying the Great Author and Finisher, we had to spin our wheels, while denying the doing so.

More information had arrived and much more so that a Hyperespheric Probe was being prepared to journey to Alpha Centauri System. That traverse would only take about six minutes at 370,000 times the speed of light. Now ground truth and in SITU exploration could better well ensue. At this time in human history, we would be able to gather more than just looking at a probe's images.

In such a Hypergeometric course, it would only take around six minutes back in time each way. With perhaps a month there, then a forward duration month, with twelve minutes reverse duration, and at 4.32 light years away, of very little consequence. With the greater distance of space and the greater reverse of time, it had been well established that kinetic interference would be almost irrelevant.

If a light signal is sent to us there from Earth, it would not arrive until 4.32 years later; likewise if we send a light signal, it won't arrive for another 4.32 years; so either way we have no intelligent signal to receive, for we shall return before it can arrive, either or both ways. Thus even with the massless photon only carrying momentum, riding the Event Horizon of our superuniverse, at this 'surface', we have no causal improprieties to deal with. God's creation is so well constructed, that for all practical purposes, it mechanically demonstrates Hawking's Causality Protection Conjecture.

At smaller distances, the time of reversal passage for information is irrelevant, per the Tunnel Diode; and for far greater distances across the cosmos of difficult to explain superluminal observations, even more irrelevant. How to qualify this irrelevance? Consider that the kinetic signal to noise ratio from such manifestations is so weak as to have such a low interference, as to be for all practice purposes of kinetic interference, irrelevant. Also this again substantiates Maxwell's Equations of the nineteenth century, implying such a far different signal strength between retarded electromagnet waves to advanced electromagnetic waves. The retarded waves which we are well able to measure and observe, are so many factors stronger than the far weaker whispers of the same emanating advanced waves. These very weak advanced waves are ineffective in causing any practical consequence to the electrons sensitive in antennas to oscillate by them. Though on a far more resolute measurement with enough cryogenic strategy, one could just about glimpse such whispers, but for the remaining background noise still somewhat inherent.

On this surface, our sun and her planets are traveling through our home Galaxy's disc, its spiral arms and open spaces between, its clouds seeded by cosmic waves, and slowly decreasing magnetic field. Go back far enough in time and space, and our Earth and the other planets would surely look like an alien 'exo-solar system'; past and future would be ever so much more difficult to discern, all looking the same, but not exactly. Which is what? At a far enough galactic time and/or space deepness, the cosmic is so very similar to the quantum. The relativistic is the transition it seems between the two. Hypergeometric Mechanics seem to fill in more of the incomplete that shall always exist, no matter how refined our resolutions of measurements.

Chapter 8

Humanity Reaping all
that it had Chosen to Sow

The city was active and vibrant. There were great stone works around, though not as articulated as those in our time would look upon ancient Greece or Rome. The ancient Egyptian's buildings were unadorned in some of the outer surfaces but inside seemed to have more hieroglyphics. Grecian and Roman grander architecture had more artistic embellishments throughout.

Perhaps, the structures that ancient Egypt built were not all by humans. Perhaps there lies a bridge or interface between what the giants had constructed, then the ancient humans inhabited, and then later built? The giants had more brutish contracts. What possible advanced technologies they had were probably rudimentary for function and not for beauty. They were, from what records reveal, not beautiful but rather ugly. They were not made in God's image. They existed, I am sure for His extended purposes for which we may never fully know. But as a judgment and testing of humankind in its chosen rebellion, they were quite appropriate.

It was morning as the newcomers awakened to not only the rising Sun, but their new situation. They were mystified at so much so soon. Not only were the food and drinks offered a bit intoxicating, so was everything around them. They had not to lift a finger, and all seemed as 'heavenly' as what little they were handed down from their ancestors. Could all this be what they had missed so much in their dim past? Or better, would it be to realize that all this now was counterfeit, to that which was once with God?

Climbing up on a ledge overlooking this city, some said built by a civilization from the sky, one of the younger ones peered through an open artifice. Before him a vista of thousands were moving along the wide streets. Large brutish guards and soldiers were always watching and directing the people going in unison one way or another, never two directions on a street way. At the end of one such corridor was a plaza and upon it a small, truncated pyramid like structure. At its apex was what looked like the 'teacher' sitting upon a large throne. Was it the same teacher or another?

The young man looked back into their new abode from the ledge and saw everyone eating food that had been delivered. They were hungry and many gratefully ate and then drank more of the fruit-like juices. They were becoming accustomed to their new life and were without restraint, enjoying its pleasures. Satiated, some fell asleep; this system of being fed and resting went on for some days, becoming habitual and anticipated. Many had questions, but with the effortlessness of all needs so quickly satisfied, acceptance slowly grew. By and by any resistance or anxiety was giving way to assimilating first to the new habits and then to a more pleasing familiarity with their hosts. They were beginning to trust and rely on them as their accessible benefactors. God was vaguely remembered, but without any substantial history of their human course from Eden, they were very vulnerable. Without a history to study or learn from, they were apt to repeat it. Their ancestors had run so far from Him who was their Creator that they had begun to depend upon themselves and anything or anyone who would fill the void of their choosing. They and their ancestors had continued such dangerous self-reliance, and now had new gods to acquaint themselves with. There was no restraint on indulgence and pleasure.

After a short time, most of them had internalized their new life. There were only a very few who wanted to return to the savannah, and some of them in time would be won over with the complacency of being taken care of by those from the sky.

But there were a few that sought to abandon this new way of living. They were offered a time to talk to one of the teachers to let their dissatisfactions be known. Those who did, never came back. It was said of them that they had 'gone back to where they had come from'. Soon the time for them would come to be of greater asset to the giants, these dark angels from the sky.

How could they understand their place amongst the stars in such deep time and space? How could they know that what was known on Earth as predator and prey was also among the stars for non-human intelligence and technology? How could they know about their ancestors' God who had already created for them much more, and in benevolence?

On such a beautiful Earth with our sun in a presently calm solar system orbiting around the perpetual zone of our Milky Way Galaxy, human history was proceeding on a course at great vulnerability to its surroundings. For self-serving in their own agendas for generations humankind had been under a far more egocentric master.

It is not uncommon for some species of insects, such as ants, to use aphids to serve their purposes. Symbiotic relations often benevolent are common amongst many higher species too. But in some instances such as mating, the insects also devour their recent mate as part of their nature. Even humans can be benign in their raising of lower creatures, some as pets, others as food and some for breeding as well.

If Earth can be considered a model for a prodigal Galactic astronaut, then we may be able to tease out a trend rather than specifics from this Galactic ecosystem we are a part of. How often in the past and future, and how near or far away are such things again repeated in the cosmic nature of God's greater creation?

That which is rare is more certain, given enough time and space … that is in the implications of the mathematics of probability. God is the Master Mathematician too, as He is Master of all that in His character He has, is and will be creating. Our so limited and arrogant puny humanity is nothing without our God. But with our God we can do that which we otherwise find impossible.

Far from the savannah and upon more hilly country was another group of more established humans. There was an older man there building a large structure of wood. He talked of using it upon the waters, but such waters were known to be very far away. He talked with great wisdom and was the father of some sons and daughters. They called him Noah. One man was far older and was Noah's great grandfather, named Enoch. Enoch when he was a bit younger had been hunting and had come to a valley where he saw strange men gathered. He took meat home to the village of his great grandson who was but a boy. Enoch had planned the next day to hunt again, but to go deeper into the valley.

These strange men were of two types. One group was more brutish with smaller, rounded heads. But there were some that were very strong and tall, ten to nearly twenty feet in height, but slow moving. The smaller ones were swifter and acted as guards, directing some activities. The others were thin and lanky, with large elongated heads that shaped back some. They were the teachers that also seemed to be able to look right into and through a human being. They were very cunning and convincing in their speech and enticements. They seemed to be able to lure one closer while at the same time being an inhuman fearful presence, hideous to the eye.

Enoch watched them and the teacher ones seemed to sense his secretly observing their activities. They would scan the brush and stare into suspected areas. He was very cautious, and then would slip away as quietly and swiftly as possible. They were intelligent predators it seemed to him. They were needing humans for some reason, and there were humans being drawn to them.

It is written that 'there is nothing new under the sun' and we might add under the many suns of our home Galaxy. Perhaps God was protecting His children from this and other dangers. We had a perfect place to grow in Eden and become all that we could be. Now we had counterfeits and many places. We could indulge ourselves in an amoral way while under a tyranny that we could have not imagined. Things appeared much easier without obeying our God. We no longer had to learn from Him, rather we had others to teach us; but they did not have our welfare in mind. They had their own self-interests and agenda to attain, and humankind was one of the means to get there.

As with other short cuts in life, this also had consequences. We had been driven out of Eden and now had been on our own for some time. God still had His people who chose to follow Him, but they did not have it easy. Often they were driven away from the tribal groups or even killed. When they did have opportunities they would share their understanding of the world around them, including warning of these dark angels, these giants.

On a planetary scale the Earth that God had given us was being invaded from beings far from our solar system. They were from Alpha Centauri and were the descendants of others before them who roamed the stars. For eons, there were amongst the stars such generations of non-human intelligences and even technologies, some similar and some different to God's human beings, the 'apple of His eye'. We were made in God's image, but they were not.

Many at our research institute struggled with the concept of non-human intelligence and technology. They forgot the many examples of insects and animals, who having a social order also used minor tools to hunt or take advantage of their environment. They seemed to wrestle with nature, God's natural world and all it implied, being more than just upon this planet Earth. They could not take the next step begging the question of life and intelligence beyond our immediate space.

Here in these transmissions from our first interstellar probe were images of ruins. They were similar to the large constructs our archaeologists have run across for decades before this and prior centuries.

But now another probe was being readied. A Hypersphere, a machine utilizing Hypergeometric means to travel great distances, but in reverse time. In it a crew of seven, mostly scientists of varying orders, to investigate our human origins and history from the nearby stars.

I requested to join them by adding my name to the list. There were many interested but many would be rejected. The stresses were far greater than using rocket propulsion, with its high G forces and comparatively slower speeds, thus taking longer periods of time to navigate to the outer planets and then to the stars. Life support alone would be hazardous requiring great redundancy. All systems had to be of such quality and durable in function as to need little or no sustenance from Earth.

As I have eluded to before, this all made me wonder of our great past, and how often Earth in its fullness must have beckoned others from far away. But only those creatures capable of enough intelligence and technology to be able to traverse the distances available to them in those times could send a probe or better themselves to our magnificent Earth.

As our sun, Earth and solar system made its great circle through our Galaxy's arms there had to be other creatures along the way that could watch from afar. There were supernovae, explosions from our Galactic core, influxes of cosmic rays, possibly viruses from other stars and planets flowing in the stellar winds and lighting upon our Earth.

If we just look ahead and behind in our present course on this great galactic orbit, we see Orion with some stars around 1,500 light years behind us; and Cygnus ahead of us with some stars near eight, and seventy and 1,400 light years away, in orbital line presently of our path. Here in just this very small sector of our compass, are many interesting stars and conditions we have been and will be traveling through, in the last several millions of years and the next millions of years to come. My colleagues and I view this as part of the great ecology of life, or possibilities of life, beyond our home planet and solar system. We all seem to inherently wonder about such things.

Our written human history is limited. Delving into prehistoric times, we have to depend on artifacts; then further back in time we move from human-made to the remains of God's natural world, animate and inanimate. Anything fossilized is limited as well, especially over long periods of time, and such minute evidence remains residual even beyond our detectability.

Even those things from the latter nineteenth to early twenty-first centuries will follow the same capacity of permanence. That which would remain after several millions of years would be the most stable. The more delicate would be gone first unless under very rare benevolent environments. If a dinosaur's feather can be ascertained in what is left behind, then perhaps a newspaper similarly protected and discovered well into a future time.

Often we contemplate 100 million years ago and are overwhelmed; but how about 100 million years from now? In that we seem far more humbled. Our limitations come home as does our mortality. We can pick up and examine that fossil of an ancient inland sea and all those shells in this textured, three dimensional piece of time. Yet all is even now as a photograph from God's camera. In 100 million years we will be as such in some other's hands and scrutinized, but whether human or otherwise, we know not.

God is not limited by our comforts of time and space. He is eternal and wonderful. We are eternal only beyond our physical manifestation and that is beyond what many of us care to contemplate. We are all traveling through not only space, but also time. One has an effect upon the other, always interacting.

Again, back to our Galactic orbit, for around sixty-two to sixty-five million years ago our solar system was in the Galactic arm,

Scutum-Centaurus, and in another sixty-two to sixty-five million years or so, we will be in the Perseus arm. This is considering a Galactic orbit of about 250 to 260 million years. In both cases are the concentration of density of stars and interstellar clouds. Perhaps in both, the one before and the one to come, were conditions that increased cosmic rays upon the local ecosystem, our living Earth.

In between the arms would not be available again until 124 to 130 million years hence, as it was about the same earlier in our orbit. On the other side of our galaxy is where it is benign as far as we know. So now we better understand where and when in our Galaxy are the more and less of habitable conditions on the interstellar and galactic scale for our Earth and its local environs.

Again, not even considering other galaxies, especially those so similar to ours, we have quite enough to ponder. This is our great ecosystem that we interface with, easily one billion years prior to and after our present time. And with around 27,000 light years as a radius, we are thus given a circumference, an orbital distance of around 170,000 light years of occupation on these scales. We have quite a stage for our history with so little data. What we do have is suggestive of that which we do not. Undoubtedly we are sensing trends for our Earth and our sun and the rest of our solar system.

It has been speculated that for primitive bacteria drifting by interstellar winds, and given technological intelligence slowly moving at possibly one-tenth the speed of light, the Galaxy should have been inhabited over some ten to easily one hundred million years ago. This panspermia, whether bacteria, or better viruses, and with the much more rare intelligent technological capable propulsion means staying with just rocket principles, is all within less than one Galactic orbit of our sun and her planets.

All this is quite heady, but fascinating, of course. I could not but wonder about the red dwarf star Proxima Centauri. So easily it is put to the side, but over the decades increasing interest had been revived in planets and environments around those dwarf M type stars.

It is understood that such stars are extremely dangerous with the flashes of ultraviolet and x-ray radiations in their sporadic flare-up episodes. There may be more where viruses and simpler bacteria might inhabit. And if such advanced creatures as we are proposing existed within the warmer Alpha Centauri A and B stars, then they would have probably also explored Proxima easily and long ago.

Brown dwarf stars and all their varieties were a whole other extraterrestrial realm of habitat for the simpler forms of life, but in their atmosphere perhaps not. More in their habitual moons, if there were any.

It takes just the right environment for an intelligence to proceed from conversation to tool making, to conquering its local environment. Subsequently to leave their home, traveling into the skies and lands of other planets, would be an enormous step; then to travel on to the stars, the local ones at first, searching for other habitats, food, and shelter. The spectrum of activities' complexities would be proportionate to the intelligence involved and its ability for technological attainment.

My philosophical and scientific contemplations so numerous and diverse, are all based on established scientific premise. I felt so sure of my theories in relation to the data from our first probe to the stars. Upon our arrival and becoming familiar with the contacts that we were observing, I was positive that we would be quite able to connect all the 'loose ends'.

By Hypersphere the journey would be fast and safe, as much as any built craft could be, to and from the planet that is now an older version of our Earth. It was a larger and drier form of our Earth that once harbored those who came from the sky to interfere with us, after we defied God. His protection no longer available, we were then very vulnerable to such predators. We were surprised that we could be sought after as an animal, as we pursued our own purposes.

The hunter became the hunted. The 'witted' becoming the outwitted; outsmarted not only by another creature, but first having outwitted ourselves. We chose our path, and were forewarned about consequences in such choosing. We did not like to be told what to do, even if it was for our better fortune. We were such a stubborn lot, and tend to still be. We choose to learn hard lessons, when so often easier ones are offered. We love to have our own way, even if that way has such severe risk to our future.

I was selected for the mission and went to the briefing room at the university. It was quite a conglomeration of people. Much more filtering was needed among the one hundred selected, as seven would be the final number of those going. It was necessary to be assured that health and safety would be primary to be able to attain the goals being set. Medical filtering of each person was first, then the particulars in the sciences. Finally, the technical side, for we had one on board as a pilot with a back-up as our co-pilot, and then each of us had to be able to survive if something would go wrong.

As old as the sea, with the ships crossing the ancient waters to someone going on a wilderness trail, each one tries to make a success of the venture for all involved. Health and safety, then expertise are surety to the completion from the simple to grand of any exploration. In the next few weeks we all would be awaiting the results of our briefing and testing.

I had so looked forward to the more tangible and clarified of how our theories would be. What of any revelations would finally bring some closure to things we humans have wondered about for ages? Here again, the future was revealing the past, the past revealing the future. We humans think of ourselves so myopically that we are also nearsighted in time and space. We are as egotistical children concerning our place in God's creation. Yes, we are special, only for His reasons and designs. We are made in His image, but we should have followed His designs and teachings. We have for so long followed other teachers and tended to other ways, which we have been unable to complete what we thought we should have accomplished. Now after all this time over the centuries, we finally had substantial evidence for what had previously happened to humanity.

Chapter 9

Pre-Human to Post-Human History

I had looked at the data of the orbits of our sun and the Alpha Centauri System as they co-orbit our Milky Way Galaxy. As best as can be determined, they are quite congruent and very close to our present position here in the Orion Spur and on the other side of the Galaxy, on the other side of our barred core. That is about 124 to 130 million years ago and the same into future time. The Alpha Centauri stars, including Proxima Centauri are more elevated from the disk than our sun's compass, but get approximately as close there, as here. In another 20,000 or so years, Alpha Centauri will be even closer. This closeness from three to five light years happens twice per the Galactic orbits of these two star systems, which are both around 250 to 260 million years per orbit. The Sun and certainly Alpha Centauri bob up and down the plane of our Galaxy too; this adds more variables to their orbital geometries.

If we just focus upon human history over the last tens of thousands of years, we see that it is quite interesting, these two in their closeness of not only their celestial mechanical dance, but in their constitutions. Alpha Centauri's stars are all quite older than the Sun and have more abundance of heavier elements. Some have speculated that unlike our present understandings, they are from different interstellar nurseries. But that is quite debatable. Being older stars and so closely rotating around the Milky Way, they could have come from a larger cloud; so the Sun is a younger of that large cloud by a billion years or more. Thus it has had less time to synthesize the heavier elements that are more in abundance in the Alpha Centauri's stars.

The co-orbit is highly suggestive of a common origin for these two stellar systems, our singular sun and the other a triple, the Alpha Centauri A, B and Proxima. This also has allowed plenty of time for life's Inherent Differentiation (IH) (that replaced the unprovable by repeatable evidence older evolutionary theory) to topologically unfold in the time and environment triggers amongst all of these stars' planets.

Given time for the predatory dark angels or giants to develop their society with technology also differentiating progressively over time, they would sooner or later seek out other fields to conquer for their own proliferation. Since the time that the protection upon the Earth and perhaps within the solar system was removed by God after the human expulsion from Eden, then it was a matter of time for the defiant to pay the price of its rebellion. To reap what was sown. Predator and prey was what humankind had won, and instead of freedom, oppression would result.

God would still answer the prayers of His ever-seeking remnant, and even guide them to safety and renewed possibilities. But their journey would not be easy. Noah already was listening well enough to begin to build something that would only bring derision from so many of his neighbors.

In such dry desert and savannah regions, why would anyone build such a very large ship? These inclinations were surfacing while his great grandfather Enoch was running to avoid the possible detection of these giants who were approaching where he was observing them.

Upon the large throne-like chair sat the teacher of their group, one of many teachers over many of the brutish guards and far more numerous workers of these visitors from the sky.

From the teacher's elongated head, his piercing eyes gazed the distance. His teeth showed from his ever parted lips in such a grimace that his jaws at times slightly grinded. The pale, sickly, grey skin of his arms showed under his toga-like covering. He raised his thin, boney, veined hand and a long finger pointed slowly to where Enoch had been spotted. His penetrating eyes were sharp, his face though so ugly and scary, mirrored a very intelligent brain. The guards that rushed after Enoch had returned but they had not caught their prey.

'H-eh-h-h' came the gurgled hissing sound from his throat. Then his so slow movements shockingly became sudden as he turned upon one of the guards, aimed a long metallic stick at him which incinerated him to plasma and ash. The heat was extremely intense from such a micro-nuclear type weapon. The other guards stood silent and motionless.

Almost inaudible to the ear, 'Ah-h-h-h!' he expelled, his eyes penetrating into those of each guard. Then his extended hand's long finger pointed to the area that Enoch exited from.

Then in a silent, yet communicating way he 'spoke' into the minds of each guard. They 'listened' as if they were hearing his voice, but to anyone not of them observing, all was intense silence. Then he seemed to repose to a quieter position, and from his mouth spittle dripped as if he was drained of his emotions. Upon his chest was a necklace with an amulet-like appendage. It lit up and he touched it. Then all was quiet and the guards left him.

In the distant horizon arose a sizable series of pyramids. Upon the plain nearby from a former visit thousands of years ago, was a slightly rain-eroded figure of a lion. Thousands of years from then Napoleon's soldiers would shoot a cannon at the head of a pharaoh that would replace the lion's head. We would call it, the Sphinx.

The giants, the dark angels, the brutish and the intelligent, needed other workers, Earth originated workers. These had to be of the highest intelligence and yet physically dexterous enough for finer constructions. They were also needed for breeding of these creatures that they were collecting for food and labor. These creatures were made in God's image and were called humans. They were social and were 'teachable'. They were 'Somebody's best', but had gone very astray.

A small distance from this city of beings from the sky, were large 'farms' with many pens of human beings being prepared for breeding and training. A fair distance from this farm was another that was out of sight from this complex. The farther establishment was for the elimination of those who were found undesirable. They were only good for food, and so that is what they became. Some were hunted for sport as well and then prepared.

From these would derive more strange amalgamations from human to subhuman to superhuman. Confusion reigned upon the Earth, humankind that remained somewhat civilized had bottlenecked into a period of a very dwindling population, and Noah had finally constructed his ark. His attempts at first were met with derision, but as rain had been looming for days some were wondering that he had been the wiser.

Such human history would unfold as we would know it. And for myself as a scientist, I was also enthralled of what was to come. From about 120,000 BC we had learned of more of our prehistory to what might be to come. I contemplated our history to be for the next 120,000 years. But to better scientifically ascertain that would require my actually going by Lorentzian effects into the future and leaving behind all I had become accustomed to. Yes, I could come back, but that would require the extreme technology still under experimentation of the Hypersphere. Dare I to venture such?

The past is as the colloquial 20/20, but the future is supposedly unknown. Yet it is only unknown until it too becomes the known past, thus history. As an old man, why shall I not venture so?

Into a post-human future would be quite a courageous undertaking, as I would then be utterly dependent upon Hypersphere technology to get back. But from the little I knew, I would only return upon a parallel world line, an alternate but real history.

I should then not only travel to Alpha Centauri in the six minutes for the entangled tunnel to allow, but also to venture a glimpse of such future as to very possibly leave humankind as only a shadow, slowly still disappearing from God's inertial recorded books. While all things never really fully have any trace after all.

When appropriate I would travel by Hypersphere into the far future, deep past and far distances in space. My, what concepts! We were talking no longer upon local space- time, but of such considerations of space and time with such great distance, that space and time become less and less defined separately. It is though at greater distance, and Bell's Non-locality, then also with greater time separation of a more quantum 'now' that is so much less defined from past and future. Travel millions of light years of space and also millions of years of time, then we have such a sameness that 'eternal now' is glimpsed.

What if I journey one million years past to one million years future? And also traverse one million or so light years away and back to origin. What then is the relevance of not only 'here and now' but 'past and future' conceptually? Are we then at the doorsteps of God? Are we well beyond our element of reference and dominion? Yes, I say very much so. I am looking at a more expanded context of one million years of time into the past and also one million years into future, but also involving distances of nearly one million years of space. Why? To have a far more cosmic theater for human history of far lesser times and space. When we pick up a rock, we need the greater context of time and space of the environment of that rock to better understand it.

All I would need is a glimpse from one million years past, and then forward to contain the 120,000 years past and forward. This would somewhat make more sense, while so easily putting us into a greater Galactic view. For is not the Milky Way Galaxy the Galactic eco-system we are a part of? God is not limited by our feeble views of time and space at all.

Shall I not now be hindered by time or space, even if that time is but a glimpse? The farther distance of time or space, the less the duration of the view for the time or space will be. Thinking this way so overwhelmingly, considering one hundred years past and forward at one hundred light years of distance seems so much more manageable, indeed.

It was some days later that I had pondered a celestial sphere of the stars that we oft see from our Earth and their placements upon the night sky. On average over one hundred years all is quite where it was, though of course for the most subtle of movement. Thusly our sky looks the same. This changes more so with time, and to ancient Roman eyes it would be ever so slightly different. To ancient Egyptian eyes more so, and going to 100,000 to one million years, the sky would begin to change much more so. Within the expanded time frames of the past and of course the future, our sky begins to appear differently, not as to what we would expect to see.

The Earth revolving around the Sun, the Sun revolving around our Galaxy and our Galaxy cutting its course through the Local Group and beyond, we have in reality an always changing sky.

As I was exploring this celestial sphere of the mid-twentieth century in the archives of our university, I noticed that to retrace the course of our small Earth through time, with its concurrent path through space, then we would have to place our world in history back towards the Orion Constellation. Considering the Sun coursing roughly at one light year for around one thousand years, then we would be about fifteen light years towards the general direction of Orion. I am simplifying such dissertation for the sake of gentle contemplation, for sure. Now that would, without counting the movement too much of our Galaxy, again place us towards Orion and give us some retracing conception of the space involved. This would be the inertial and kinetic in forward time, but just considering back to around 15,000 years or so, to about fifteen light years in rough calculation.

In order to traverse this at 370,000 times the speed of light with the Non-inertial chord through Hyperspace, it would take about twenty-two minutes. But because of the Hypergeometric Mechanics involved, we would be fifteen light years away, but in the negative twenty-two minutes. Space would surely be traveled but only with a slight change of time of just fifteen years. To traverse nearly 1,500 years of time, we would traverse around 1,500 light years of space, with a traverse in tunnel time of thirty-six to thirty-seven hours. For in the tunnel it is forward time, but to anyone measuring from outside the tunnel it would be negative thirty-six to thirty-seven hours. And even more counterintuitive than all this is already, we would be 1,500 years in the past.

Now to find our Earth with enough precision to place the tunnel and entanglement point to travel to is daunting. One is traversing with the predominance in time to a reactive element of space through a chord in Hyperspace, 1,500 light years away, not fifteen light years away. For now one is actually Non-inertially retracing their world line to a past point. The spatial past point would be located at 1,500 light years away, when our Earth was at that time, while the spatial inertial would be fifteen light years away. But that spatial was not considering the summation of movements of not only the Earth, but the Sun, our Galaxy and even our Local Group, and the rotation of our universe floating upon the surface, also in rotation of our superuniverse.

All so counterintuitive only if we maintain an inertial and kinetic perspective within our celestial mechanics, disavowing the consideration of the Non-inertial and irrational mathematics involved with reverse time dilation.

In the archives was an old paper from around 2010 reporting the experiments of those just one hundred ten years ago. This was in honor of H. G. Wells and his work, *The Time Machine* and brought this temporal voyager of such Hypergeometric Mechanics to 1895.

That distance was 1,500 light years and not the assumed approximate almost one tenth of a light year one would try to ascertain with the Sun's one light per one thousand years' rate of revolution.

Such is the degree of the counterintuitive in Hypergeometric Mechanics, surpassing that which is inherent in relativistic physics and quantum physics. These sciences have their limited perspective of inertial and kinetic extremes, because of the supposition that nothing material can equal or surpass the speed of light.

The purpose for my speculation upon 1,500 years was to try to glimpse some of ancient Rome, perhaps with a robotic probe. It would be less conspicuous if it were to fly over ancient Rome, be quick and hopefully unnoticed. Perhaps the Hypersphere should remain in an orbit while dispersing an atmospheric probe to acquisition the lower atmosphere. Such a probe would be wisely colored to blend with the sky overhead to remain as passive and non-interfering to people of the City of Rome in around 600 AD. A distance of 1,800 light years would place this endeavor at around 300 AD and so would possibly be a better time to view a more typical period of Rome before it had begun any perceivable dissipation.

Over and over I would calculate and surmise all these things. To go farther back would also require such camouflage of mechanism so as to continue to not impede the sequencing of the flow of events as we try to know them. Consider 120,000 years and then one million years, what windows upon the past would be exposed? And to forward transit keeping within relativity's mechanics of inertial and kinetic forward telescoping of time is also of grand vistas to witness.

Pre-human and post-human history enough to glimpse, thus sets in context the story of humankind ... all spatially from our Earth to the stars, and temporally from our comfortable present, to times deep past and deep forward.

Such symphonies of time, then space, were grander to ponder. Yet I had to maintain an even keel and keep scientific while exploring these far longer histories within histories of the origin of humankind. Yet, with simplicity, there was a common denomination. There was a phenomenal origin for a phenomenal creature, called human. And all this other scenery was supportive of such a grand story.

From the far past to the far future, and from far away in space to an opposing far away in space, was still now and here. Resultant with increasing resolution to the most detailed of the present and this Earth.

While pondering, I had fallen asleep. Then in the early morning a call had come in. There were more details from the mother probe orbiting and the smaller probes that flew down to the surface to examine and take images.

From the congruencies of the ruins of the Earth-like planet in Alpha Centauri's, one of its two major stars were the same age as the oldest ruins remaining upon our Earth. These on our planet are some of the oldest of long disappeared civilizations. Sumerian and Egypt particularly, but also Aztec and Mayan were considered as such.

These remnants of the dark angels, the fallen ones, had endured for us in these times to better understand our origins and our story. What was to come had arrived. A better world was here, where it all made sense.

The dichotomy between religion and science had with the passage of time, been solved by the long awaited arrival of the One. A New World for humankind had come and all of us in our present day had been living in a Millennial Epoch. Humanity had throughout the time of human government so failed and limited itself to a mere shadow of what it could have accomplished in obedience to its original Master and Creator-God.

Human history had so repeatedly demonstrated that humans are really not capable of governing themselves, at least progressively. We tend to feudally organize and govern as a hierarchy of masters over slaves, no matter how benevolent we try to be. In time we corrupt and disintegrate to our baser natures and thus digress to some form of violent revolution that perpetuates to refinement, then corruption. Yes, there had been technological progress, but such knowledge and aptitude had always seemed to leave the appropriate wisdom behind. So we continued, primitive-feudal to high-tech feudal, with a time of cruel and arrogant Rome in-between. This seems to be what summation I was left to succinctly consider.

Before the Fall and after the arrival of the One, humankind returned to its prime state of excellence. With the phased return it was a time for restoring what we were designed to be, as like the phased descent from our ideal status. The great mystery of the far past and for some of those in-between a far future, was again enlightenment of what was for a time beyond our grasp. We were not allowed to go as far as we could or to understand such, for our hearts were not aligned with His Will. It would be as if a teen-ager were given the keys to a race car but lacked the wisdom to really handle it.

We surely had the potential but continually failed to attain. Humanity in such rebellion was basically cruel, unfair, unethical and self-destructive.

He had to come again in order for us to return to our full aptitudes. Without such seasons of defiance we would have been to the stars ages ago, along with such arts as we cannot imagine.

Those of the human family that had chosen to submit and surrender to their King of time and space knew now the fuller fruition of all that was promised in the ages past.

As Saint Augustine had said, "God is the maker of worlds," and I had always added, 'and the destroyer of such'. In reference to these wonderful times I have added, 'the maker of renewed worlds'.

And what of a post-human history; where may we be going? Perhaps just beyond this Earth, where the One awaits, some of us may go. Perhaps we are to expand amongst the stars and also occupy this Earth. God's greater ways, as He revealed in His word in Isaiah chapter fifty-five, may continue to insure our rational limitations.

Let us mildly contemplate, even after this Millennium of Peace, that beyond we have at least 120,000 years. We have basically looked back the same amount of time, but history is better viewed when it has some evidence of what has happened. All we can humanly do is extrapolate based upon probable trends. The past revealing the future, albeit in a limited way. Yet we still see some things through a glass darkly, as others have been made far clearer.

Even H. G. Wells in *The Time Machine* envisioned into the future, especially of the year 802,701 AD. What a daring concept indeed. Subsequently with a furtherance of humankind after the Millennium and the One upon this Earth, there are still some not wanting to follow Him. So in this context, a Wellsian view is a fair possible order of what could come.

For in 802,701 AD humans have digressed again without God and are left for generations to develop without Divine intervention, as their fore parents had chosen. By this time only derivatives of what humanity was before under God, but also of two subsets separated from God: one mechanistic in technology yet brutish, hunting the second subset that was non-technological and childlike. Predator versus prey. Please bear with me for my interpretations are but that, mere interpretations.

If going to one half of a million years into the future is such a stretch, even going 120,000 years into such a time to come is still quite a strain upon intellectual license. So with the benefit of a lesser distance temporally, could we probabilistically contemplate the future of humanity?

Through 120,000 years we have had some ice ages, the rise and fall of oceans and climate variations; so too this should continue, though with some influence waning after the passage of our industrial and technological epoch upon Earth's continued history.

One point I would like to make, again in reference to Mr. Wells' great book where he illustrates the 'time traveler' in a period of time when the Sun had expanded to about one third of the sky and the Earth had slowed to almost stationary in its rotation. This may well illustrate a time when the Sun had expanded to a red giant in its dying days and had such devastating effect upon our Earth. That surely is considered by natural science to be quite the scheme of things by that time. Many estimate around 500 million years more for complex life as we know it to continue. Prior to 500 million years ago, life was far more simple and the Earth not so benign to larger scale life forms. Thus in another 500 million years the Earth would render severe conditions for advanced life forms to inhabit. So as the Sun's increasing heat continued, only simpler bacterial and viral life forms would eventually try eking out any existence in the harsh environments. Where once oceans and atmosphere were sustained, sadly their volumes of water and air would be dwindling. The increased solar winds and radiation now impinging upon the vulnerable, remaining, isolated abodes of the last strains of life, meant slow death on this Earth's surface. Perhaps, the subterranean world would continue for a time until the rising heat eventually vaporized Earth; and/or the oceans of Europa and other water-bearing moons would suffice with their similar Cambrian explosive analogues.

Everything has time and space windows of possibilities it seems, and so does the present, as past and future states do, of humanity. Rare and benign conditions occur infrequently and so distantly as to make the local space-time so lonely to any of God's creations with intelligence, human or non-human. 'Like' seeks, and understands like, whether in a small town or at a young age, or light years between the stars, in a separate yet similar way of what we perceive as the present or now.

Chapter 10

Reflections upon these Greater Scales

It is of utmost importance, having had such a grand meal, to now digest slowly and let all come into some order of consideration. These things we learn and think are so much like that. First the taking in and enjoying the variety and flavors, some new and others familiar, and some with just acceptance, but difficult to desire; then with a glass of wine and a calm place to relax, the reflections slowly surface and make a wonderful experience of contemplation and conversation.

Like one taking a holiday to some Caribbean islands on a cruise ship and then returning to family and friends to share. One is now back home and yet the journey continues with fond memories and interesting adventures. It is in the sharing and repose of the familiar again, that the absorption and reflection are enhanced.

So too with our far more expanded prospective of insight from the Alpha Centauri Probe and the following up with the Hypersphere Probe. Yet with greater discoveries, are greater questions.

Now with such revelations blessing me, I so much more accept my limitations and let the grand mysteries go on. I have had such opportunity given me and can see farther than before, and yet the wonder of it all continues. I still see through a glass darkly where my clarity of vision ends, and I think it always shall be. God's great designs within designs are so wondrous and sometimes it is just so magnificent to savor it as the limited creations that we are.

From the minute to the enormous, from the very slow to the very fast and from the distant past to the near future, all is of scale, from Planck this to Planck that. Resolution is the fine comb we use to discern and it is the limit of such resolution that we are brought from vision so keenly sharp to further vistas of obscurity, so darkly clouded.

It is accordingly for humans to wonder of past and of future. Myopic is the present moment, so we think just here and now. But there are trends and events that have happened in the past, as there are events to come in the future. Such rate of flow is relative to other rates of flow, so there is no simultaneity based on the local curvature of space. This local curvature is the consequence of what is popularly called 'gravity, which had by our time been redefined as Inertial Geometry.

Take this lack of simultaneity far enough into space, with approach or recession between two comparative points in space. Then increasingly distant will be the one in the future to the one in the past for their respective nows. It is the one moving faster that has the most divergence to the other's less relative velocity.

Hmm, 'contemplations a la reflections' do tend to flourish in the more counterintuitive. But it is basically a Non-linear universe we are in, with these sub-light, tardyon incoherent areas we consider normal.

Well, this wine sure is nice to sip, and it is so nice to enjoy this summer evening with the whisper of refreshing breeze every now and then. Ah, I see the Moon in the sky, so quiet and so near, familiar and alien.

I have thought of taking a holiday in the northern plains of Copernicus, the crater in the western of the mid latitudes easily seen from Earth. A friend sent images of the Sun rising and setting over the eastern, then western rim of the great ninety kilometer wide crater. They were beautiful with the walls' reflection of the Sun's bright light and the utterly black, starry sky behind the Sun, while the immensity shined from the opposing in places. In these northern plains were caves and holes long ago discovered and now able to be explored by astronauts. Many have gone on to Mars, but I prefer the shorter trip to the Moon in my older years. As a young man, the Mars' journey was wonderful too. But it is quite the beautiful repose to look up at the amazing Earth above one in the lunar sky. It remains in one place, yet turns showing phases, and behind it the course of the constellations' tread slowly through the Earth-moon year about the Sun.

From Mars one can see the Earth as a small, almost blue dot near the Sun in the morning and evening sky. Blotting out the Sun above, craning your neck and squinting, one can see the Earth in the violet black at the zenith of the Martian sky during the mid-day. It is humbling to perceive all of humankind's history as a minor speck in the cosmos.

Fossils have been found on Mars and so a paleontological lithography has been fascinatingly revealing life in Mars' geological history. The proliferation of life in the oceans of Europa, the moon of Jupiter and Enceladus, Saturn's moon, has also revealed microbiology. Even larger scale plants and animals were discovered within the alien waters of Europa. All this indicates God's natural world extending throughout the known universe, at least amongst comparable areas of the spiral arms of such similar galaxies to our own. Within the interstellar context of the exoplanets, other Earth-like terrestrial planets possibly existed with at first chlorophyll and then other bio-signatures in the infrared. But later Hypergeometric Probes acquired images and data sets lush with such mega flora and mega fauna.

Deep space and deep time, God surely is limitless and has with repeated evidence never been restrained by any left or right political narrow-mindedness.

Such quiet, personal contemplations are so rewarding to me in my autumn years. Some of my colleagues may still be struggling to discover as much. But they are needing to discover with their own efforts instead of depending entirely on the knowledge and evidential repetitions of present scientific thought that fosters their own biases and selfish motives. This limits the divine truth, the real objective and wonderful creation of God's galactic and universal ecosystems within ecosystems.

Often I turn from space to time and then from time to space, for the two are so interrelated. Space and its causality constraints are needed for the reflected directions of space on each side of the Event Horizon. To cross this Event Horizon restrains anything to that flow of time and entropy of that side, in opposition to the other side. I so enjoy the conceptualization and then reflection upon these things.

In one text I had read that according to our archives there was a journey in 2010 AD back to 320 AD, but due to an apparent problematic operation, it actually returned from 1895 AD. This incident thus prompted an entanglement to not only that time, but to ancient Rome. I understood that it was brief, but still word of it had gotten around from the individuals involved, then another attempt had been made.

Ancient Rome in 320 AD ... what an experience that must have been ... to observe a past time in history as an actual visitation, then to extrapolate all that we have in these modern times from archaeology. Fascinating!

The discouraging thing of it all is that reverse time requires much more energy than forward time, with the same value of temporal dilation, according to Lorentz-Fitzgerald Contractions. Succinctly it is tough to actually pass through the Event Horizon of our superuniverse. Such local articulations are dependent upon the mass and size of the object passing through proportional to the energy required.

Thus for the micro-cosmos not much energy to consider, but to the macro and then cosmic, the energies go from immense to prohibitive to humanly impossible.

My wine glass waned of its red elixir and I drifted to thoughts of even farther back and forward now, to the Eocene Period, around thirty-three million years ago. Then to the future possibly another repeated Eocene Period, where the Earth's temperatures were greater and it was more humid even into the Polar Regions. I called it, a 'Neo-Eocene Period' in another thirty-three or so million years.

Now within this window the closer we are to our own now or present, the greater the resolution of the detail of available data to reconsider. Merely going back and forward 120,000 years allows a somewhat more relevant appreciation of human history, archaeology and anthropology. Now consider within a series of Ice Ages and we have a geological context to Biblical history of humankind. Was there more than one occurrence of civilization to the point of similar, though not exact advanced technology?

Looking at a Galactic time and space context for the realm of humanity and its history also expands the stage of possibilities in such. Here greater evidence of God's natural world, all of the beings, human and non-human, intelligent, instinctive and reflexive, even from simple tool use to advanced technologies, become appreciable.

The greater the space and time, then the greater the possible becomes more certain. Thus the 'what if' loses the 'if' and becomes the realization of 'what'!

Though the Moon has long ago been explored, landed on and colonized it still has a very particular emphasis for humanity. So near, yet so magnificent while familiar and alien, it has been witness to the history of the Earth for eons.

Though the base is within the craters Copernicus, it is interesting to travel to the historic sites like Mare Imbrium (where the Lunakhod 1 and Luna 14 landed) and Le Monnier where Lunakhod 2 landed. Of course, Mare Tranquilitas and the other Apollo landing sites are of the utmost importance too, all these on the near side.

On the far side are great places too where it is better to listen to deep space radio emissions where the Earth is blocked out in order to receive much stronger signals. But only the near side offers the great view of beautiful Earth in perspective to God's great creation within the vastness of space. It was here that one of the Lunar colonists had an experience which he related to me some time ago.

He had felt religious persecution for quite some time prior to this millennial time we were experiencing now. Religious persecution? Yes, for in the prior, supposedly enlightened, rational period, which was a technological rendition of how uncivilized humanity had become, he had traveled to the Moon.

Upon escaping the colonial and scientific work he had been assigned to, he waited for a field trip to our natural satellite. Purposely, he had gotten away from the group and found solace within a more private quarter of Mare Crisium. There in the sky, just below midway above the horizon's south-eastern mountains, was the beautiful Earth hanging as it had hung for eons. He related how he turned off his radio and could only hear his respirator functioning on low. He carefully got upon his knees and raised his arms upward.

"Finally, finally ... I could pray without any Christo-phobic, God-phobic or hetero-phobic intimidation. And as many had learned to do in the Soviet Union during the Cold War, I prayed!" he emotionally confided. Then he spoke more clearly, remembering that now he did not have to cower to a tyranny that falsely preached love and tolerance for all.

He surely had my attention. Here in 'magnificent desolation' he was completely allowing himself to feel his religious experience as he desired from within, but away from the Earth in the final turmoil of human governments. No one could hurt him here on the Moon, for his radio was off and the orbiting satellites and any in-lunar-synchronous orbit were not recognizing anything inappropriate in his behavior that was problematic to the enforced social-political order of those days.

Now with our great time of Millennial Peace, the flourishing of long thwarted talents and gifts, and the respect and encouragement accorded to such persons, we had made it to the stars. We had finally returned to what Eden was and now even more in our 'New Time of the One'.

What was written had come to pass in God's time and space. It was recorded that in the end, the sun and moon would turn red. And yes with our better resolution upon the Sun's spectrum, we did notice helium rising more than anticipated. The diameter of the Sun had increased some and its output of heat rose proportionally. These increases had warmed the Earth, in those days of less understanding called global warming, but reconsidered based not just on carbon dioxide build up. Rather methane (four times as influential as carbon dioxide) and water vapor had built up in the warmer atmosphere of the Earth with an enlarging and redder Sun. These changes in the Sun were not so obvious to the common eye, but more so to scientific measuring.

Strain from the Sun's waning tidal effect upon the Earth had allowed internal pressures to try to balance, and the large, prehistoric mantle plume-driven super volcanoes to explode. Krakatoa of the latter 1880s was very powerful. But the Yellowstone super volcano occurring shortly before the Millennial Era, rent far more damage. The present, great time of peace and wisdom was also preceded by other super volcanoes on this Earth. The one under the Hawaiian Islands and others also screamed out their long silent power from the outer core of the Earth to its surface.

While only waiting and watching for the great asteroid or comet impactor to come, humankind's technology had no ability to circumvent the Earth's own destructive powers from within. With the Sun's advanced aging, because in the past we had underestimated its age upon the main sequence, we were not prepared to face such a solar change so soon.

On the Earth there returned a less benign environment as was the norm before the arrival of the humans above the animals not created in His Image. The oceans roared and hurricanes of immensely greater forces reigned at times.

Into our solar system, as had happened in times past, a rare interstellar comet coursed and drove itself into the large planet Jupiter. The great red spot had waned some but another was evoked, three times larger than the one well known. It was as large in length along the northern hemisphere's mid parallel and far more compressed in its vertical measure. It appeared a deeper red after the impact which added more mass to Jupiter. On the opposite side another antipode darkening ensued and Jupiter seemed to glow eerily from its night side as well, as probes in the outer solar system relayed back to Earth.

For some time Jupiter seemed to act more like a brown dwarf and so we reconsidered the solar system as a binary one, where Jupiter was regulated to a lower stellar status.

Combined with the growing, warming Sun and Jupiter emitting more heat than previously (three times or more than it received from the Sun), the Jovian moon Europa developed oceans with waves and an increasing atmosphere. It was temporary cosmically because the atmosphere slowly dissipated away with the low gravitational field of this moon. Mercury seemed to be the core of a prehistoric near-sun gas giant, which lost all of its crust and most of its mantle, leaving only the extended core, with a skin of lower mantle remaining.

To humans a far less sedentary and comfortable environment existed within the solar system. The Sun had slowly deepened inside a denser portion of the Galactic arms as well. Cosmic ray counts went up as what must have been an area of ancient supernovae was interceded. Earth's cloud cover increased as well as the cloud covers of other planets and moons of our solar system.

These things are samples of all that God had done, is doing and will do as He wills. For upon these grander scales all with some similar trends, yet each as individual as adequate, His creation has continuum.

The season given to humanity was of sufficient time and space as to achieve the goals of His great design of designs within designs, as His wheels within wheels. Humanity's shortsightedness in time and space for each generation was a chosen limit by those who preferred their usurping God's place. It is written that 'pride goeth before a fall', as it had repeatedly been of such demonstration.

During the grand course of human events (that Non-linear sequential flow of moments upon the linear world line among parallel world lines of time-space) travel the 'nows' we often think we possess. Our myopia is such as to allow us to better live and work within such practical framework, to survive and proliferate our kind. Beyond that, for the mellow reflective of us who dare to observe and measure outside of these limited perceptions, one is thus able to soar to profound ranges that provoke grander thoughts and views.

Humanity's problem, because of our desire to think ourselves greater than we are in relation to God, was dealing with the humility of it all; and then on the other hand the grand wonder and thrill of it all. This was a problem, for within each of us is some desire to be in charge of things so we can have the advantage. We are here presently but temporary residents unto eternity. Where we came from and where we are going, we can only surmise through the proverbial glass darkly. We have no input to our origins, yet we have responsibility for our destinies. Each of us can choose to some extent our furtherance in decision making and thus the consequences.

These awesome perceptions on grander scales ease the understanding of time and space, and the cosmic scale to the macrocosmic and quantum scales. There is self-consistency per world line that is thus insuring the proper causality without paradoxes. With the primary focus on time, the resultant space travel of great distances was solved.

In this grand new time of finally being released from the hindrances of governing ourselves by the One, we could finally perceive what in our past was so often hidden by human attempts to achieve the mathematical limit of 'self-government'. We were never created to govern ourselves, but were made to submit to our Creator only. All else was a demonstration of such a truth.

I would revel in and often share my reflections with those of similar thought when we got together for discussions. It was frequently that looking into the night sky or relaxing in the quiet of the day sky that one had such moments of contemplation. I again yearned to return to space and regard the Earth and Moon, separated from them yet invisibly connected to them by the interactive Inertial Geometric in action of celestial mechanics.

My name still on the dwindling list of possible candidates, I waited patiently in my work. Just thinking how the Earth has beckoned across interstellar space for eons of bio-signatures, as the Sun and its retina of planets encircled our home Galaxy, every cosmic year, around 250 to 275 million years.

There were the strange 'natural' nuclear reactors discovered in Gabon, West Africa in the twentieth century. They were reckoned to be about 1.7 billion years old. Perhaps before the cataclysms to come, these reactors were more proof along with the large Boskop skulls found nearby, that giants of old had settled or colonized the human world God had protected in Eden before we defied Him. Following our Edenic expulsion, we had gone into a world exposed to other non-human intelligences who even utilized advanced technology. How rampant I asked myself was such amongst the stars?

Alpha Centauri A and B, those binaries nearby beguiled my thoughts. Perhaps a human journey there would be plausible. Even with Hypergeometric Mechanics, the logistics begged for redundant and complex support systems. It would not just be a day trip, though possible so easily; it would be an expedition of grandeur. Such a journey would involve great preparation ... consideration of conditions known and unknown there, and preparations for unexpected events.

In the next several days, I brought up my suggestion to some of my colleagues. Such a profound exploration of our origins was my excitement without bounds. We had some meetings with others who shared my interest, and out of around thirty people there were about a dozen who would be willing to attempt this adventure. There were discussions and brainstorming that helped so much to banter around the most essential means necessary as well as most fascinating. It would be revealing more information upon our origins.

With Hypergeometric Mechanics, great distances were now easily accessible. It was the small confines of even a fairly large Hypersphere that was restrictive. Within a few weeks we would know who would be able to go, when our mission would begin and its duration within the Alpha Centauri System.

Now bravely we would be able to actually enter into the ruins there and get whatever samples we could, but the information visual in multi spectrum would be immense. Much of that could be transmitted either through the entanglement or just by a blue laser back to Earth, if some power restriction entailed. We would spend about a week and then return in another approximate six minutes. If the blue laser information was needed, then the reception of that would be another 4.32 years.

The limitation of the speed of light and divergence of the laser beam were the added parameters to concern us using just the blue laser. It would be a good backup. Actually with the intense memory aboard and with our instrumentation it was better to just all come back through the Hypergeometric Chord between the time there, then and the now to come of the expedition. The best of all is that during entanglement the transmission of such information would be the best.

Chapter 11

Preparations for Alpha Centauri

In the southern hemisphere skies near the Southern Cross the Alpha Centauri System beckons. The distance of 4.32 light years was daunting to thermal reaction engines solely based upon inertial restraints. In the early twenty-first century the advent of Hypergeometric Mechanics and the release of mechanisms from inertial limitations, then allowed starting with diamagnetic particles, on to diamagnetic hollow spheres as payload carriers, to traverse great distances in space; but in reverse time at 370,000 times the speed of light.

It was the Precambrian Era, Super Eon in which 7/8 of the Earth's and our solar system's history had long remained mysterious, even until recently.

Given from about 4.6 billion years to approximately 540 million years ago our solar system orbited our Galaxy around eighteen times with only the minutest of single cell lifeforms as predominate. Though many specimens formed colonies and some even existed between oceans of salt water, shallow seas and low lying land, others must have been deep in the ocean or on dryer ground of inner volcanic islands to even in the clouds.

With an approximate orbit time of 250 to 275 million years per Galactic orbit of our sun and its planets, it must have at some point gained the interest of far older non-human sentient and technological life forms, perhaps some of the angels in the second heaven, space; some benevolent and others malevolent, and perhaps among them the builders or giants of Alpha Centauri. Perhaps the colonizers of Alpha Centauri came from farther away, with a much longer history.

The Alpha Centauri stars are around one billion years older than our sun and its planets. If we all came from the same open cluster, it would be possible, as compared to a more compact cluster, to have a much wider expanse of age differences. Other clusters, even older and somewhat closer to the nine parts of our Galaxy, would tend to be older and also amongst the disc.

Ponder all this with a sporadic active galactic nucleus, especially when the Galaxy was younger, the density of the spiral arms when traversed. Then we have quite the variability of our solar system's time of journeys, many in its orbit about this same nucleus. There may have been moments in our orbit when it modified over time. So give or take such relative errors, the plus or minus I have afore mentioned, and we have some very interesting thoughts to digest.

Upon this expedition we shall traverse in this future time to relevance of information so entwined with the Precambrian. The farther in space we go, even with Hypergeometric Mechanics, the farther back in time are we in more intimate relation.

Reconsidering that during the Cambrian Explosion, a mere 500 million years or so ago, where God's Inherent Differentiation (IH) unfolded the topological bio-geometries required to survive and thrive. With only two triggers, one of time or duration of repeated stimuli, and the other of the specific stimuli of environment sustained, the life unfolded and was able to adapt to all that was programmed into it. This encoding by God the Programmer, is for all life in the universe. It is software driving hardware and hardware feeding back upon software ... all that is within its genetics comes out in form and function as required.

Proliferation is mathematical, as it has such an adequate, all-encompassing fundamental of biological economy able to withstand the technical or probabilistic vagaries for that balance. Merely surviving to liberally thriving would result, unless something, so overwhelmed, would succumb to extinction. These mathematics were functioning for eons before within our Galaxy and others by God. It is only our being so near the front stage of God's Earth that we so myopically perceive His work just limited to here.

We had to study much archaeology, anthropology and even paleontology, for so much of the Precambrian was more relevant possibly to our human origins than ever considered through most of human history. This also had us reconsider Martian and Lunar anomalies, along with even the intrigues of oddities in or near Earth's orbit, like the 1991 vg anaomoly.

'Post-Cambrian', yes let's call it the 'Post-Cambrian', offered our consensus upon the far past; for all practical purposes nearly equivalent to the far future palentologically. In between we have a more recognizable view of the anthropological past and future that we have a better glimpse of by extrapolation. Then we have history repeating itself on our grander scale of civilization and archaeology. What if we consider, as our contingent of adventurers shared, that up to a world-wide catastrophe all this unfolds to a time like ours ... time of Edenic and Millennial climax of the highest of what God and His humanity can cooperate together to do?

If taking this to significance of rhetoric, then with every time of the past, there is a future time in the cycle to prefix with the term 'post'. Consider please such speculations, if you will.

So we thought that a great cycle seemed to be implied from a Precambrian to a Post-Cambrian, when within lie the lesser of the time wheels, such as Paleontological to Post-paleontological; Anthropological to Post-anthropological, Archaeological to Post-archaeological, and Prehistoric to Post-historic. Yes, this rise of the human condition from a catastrophic event or series of events, after a climax of time, humans and God could finally achieve working together. Then to a devastation of loss, the humans are again destined to find out for themselves what they chose to not take care of. This cycle of behavior might be something of interest in our human, meager, foolish attempt to understand our God and His ways.

"Ah, very interesting," said I with a smile. So possibly what we were doing was glimpsing the past and future of places within our Galaxy and others, on such galactic time-space scales where God is not limited by human perceptions and thoughts

"We are really, even in our local interstellar scale, just a larger sample of something so much bigger than whatever scale we attempt to measure," offered another of our group.

"On toward Alpha Centauri!" arose the call from one of our later meetings, with enthusiasm afore only with prudent and detailed academic attitudes. We were like children now with a new adventure and a new toy that we had heard about, but had never ourselves really experienced.

This Hypersphere was quite the vehicle for such long distant travels, but in negative time. That was not problematic, for the results would replace any concerns with great reward.

The difficulty with many things so profound is our very limited human experiences, even as scientifically we are able to prepare, despite the infiltrations of our subtlest biases. Please consider that we are so wanting our understandings to be convenient. The long cycles of God in decades and thousands, to millions and billions of years motivate our need for quick confirmations of our opinions, though we formed them with our very limited scientific objectiveness. Rather we love with our accredited affiliations to succor solid foundations for what we believe over what is inconveniently, repeatedly evident.

Theories are often promulgated as facts to the point that any rational offense causing the atheist or agnostic academic to defend themselves. They have continually thus attacked the God-believing academic and put them to such repeated defense. They have trained their prey well. But, as the Russian proverbs relates, "The hunted has become the hunter." So when chased, the former hunters, now prey, are not prepared to support their consensus of opinions no matter how arrogant their primary presentation. On the defense of their atheistic or agnostic anti-God religions, while stoutly denying such, they are found without scientific argument, and retreat from the repeatable, verifiable scientific evidence now before them.

In this 'New Time' we are in, such wasting of efforts has been eliminated. We are not hindered by Hypergeometric Mechanic's more profound potentials. There is currently politicized academia and corporate-business elites that redefine what society should be to serve their purposes. Now the One is here who has freed us and allowed us to return to what Eden was.

In this New Time we are still discovering more of what so long was denied civilization while humans proved for centuries how they could never really govern themselves. Now with our guided potentials we far more wisely are able to glimpse beyond our design limits and are permitted in such cases safe extensions to any 'Pre-New Time' limits.

It would be days until our great expedition to nearby Alpha Centauri's three stars. Yes, all three were scheduled for far finer analysis. The stars had been for so long encasing the known and supposed human history, yet so close in space as to have the most probable effect of anyone or anything of a non-human technological intelligence from an extraterrestrial nature.

Ha Atik Yomin was now upon our Earth and our society was again at its Edenic peak. This time perhaps we would not allow ourselves to be enticed by the allure of our wills and desire over our obedience to our Creator.

I had been putting together my perusal effects and reported to an orientation center for our journey. It was to be sponsored by our university. In the briefing and following tour of the Hypersphere, we soon were apportioned our particulars of its performance and a look inside.

I had wondered what it would be like within its interiors after having read and seen images. But to actually look inside something so small, engineered to go so fast, and to take us to Alpha Centauri in around six minutes, was incredible. Such a simplistic yet functional design, while on the other hand the geometries imposed upon its structure and function were generating amazement.

This one was built to hold seven to eight people. Others were larger, but we did not feel we needed that size. As this was a journey of approximately six minutes one way and around twelve minutes round trip, we felt that it would be important to leave room for any instrumentation needed. This type would require some equipment for recording and measurement, also ground sonar and radar in easily carried sizes. First aid was important as well, but that would be always within the Hypersphere's payload.

As this was a dead planet in the sense of no evidence of anyone or anything humanoid living presently, it was not necessary to be prepared to blend into any extant cultures of the particular space and time.

We had also considered that we would be in ruins of a very past time. The builders of such were the ones that Enoch himself may have dealt with in early human pre-history. We were having the divine privilege to visit, observe, measure and interpret first hand their works as they had remained for ages.

All was in order; it was now hours until our encapsulation and Hypergeometric oscillations that would allow our reverse time spatial transit. We were ready!

Chapter 12

On Toward Alpha Centauri and the Ruins Predating Eden

Inside we had secured ourselves in our individualized, horizontal couches. Each of us had only the minimum of observation through small portholes on the opposing curved wall. A slight hum ensued and we all were soon sensing the Lorentzian and Para-Lorentzian effects of approaching the Universal Event Horizon, as we slipped through its local construct. We first experienced darkness, then a blinding light that formed a ring outside, then thinned and re-expanded. At the apex of the thin, bright ring there was a dim blackness, too instantaneous for even our high resolutions of measurements.

Outside we could see a dusty, windblown landscape which seemed so distorted from the Hypersphere's transitional geometry of expansion. The same occurred during our beginning of this trek right before the ring formed, as we distorted to a compressed disk-like form observable to any outsiders. In our arrival the same distortion would be seen from any outsiders as well, but in a reverse geometric progression of a thin disk to expansion to a sphere. On our way from Earth -- sphere to disk to fade away; and then here on arrival, fade-in to disk to sphere ... one geometric function only the operational reverse of the other. Such is this chordal geometric journey from one part of normal space-time to the other through the Event Horizon of our superuniverse.

As we gathered ourselves conceptually, then physically with our equipment, we also had to review with each other our ground strategies. First we were excited that we had successfully landed amongst what appeared to be either ruins or articulated-like geological sculpted formations. We used an observation pod to overlook from above our surroundings. The air here was thinner with about 15% oxygen and 60% nitrogen, with the complement of the atmosphere about 25% carbon dioxide. In the mix were some less pronounced noble and un-noble gases. The atmospheric pressure was about equivalent to being in the Andes Mountains on Earth so we needed pressurized oxygen and special generators. The air was much drier than what we were used to as it was desert-like and again like the Andes with a chill to the air. The Inertial Geometric, gravity, was about 80% of Earth's.

We emerged and soon saw the two suns, A and B of Alpha Centauri. In the night we would try to locate Proxima Centauri, the M-type red dwarf of the three. It was similar to our G-type sun, but about a billion years older as ascertained for all three stars. B was an orange K-type also having some planetary company near it. For A we were over *1au* from A, but had other planetary companions inward. It appeared (this understood with better analysis over time) that the large terrestrial planet nearest to A was the core of a long extinct gas giant, similar to what Mercury had once been.

The winds kicked up but we had protected our eyes with transparent helmets that also guarded our ears and faces. Because the air was thin, the winds were swift and the sand particles stung any exposed skin on our arms and legs. We should have worn long sleeved shirts and trousers. These winds were sporadic and so at times the air remained still. When we were near any geological or archaeological abutments, we would be shielded from such annoyances.

The closer we came toward the nearest abutment rising from the ground, we saw that it was more like a pyramidal construct than mountains. The heights were very comparable to the pyramid-ruins of Neo-Egypt and Neo-Mexico. To attempt to climb this one, such as we were would tax our internal oxygen generators more than warranted. We instead searched along each side for any indications of a doorway and we found one. The height of the doorway was around five and one half yards. Was this the size of the builders who were particular to this ruin?

I also wondered at times if there was more than one breed or species of such giants, dark angels or builders. As with insects, there are different types of ants or bees that make similar but specifically different nests or hives.

Here we had a great edifice and evidence of non-human intelligence and use of technology. As the winds abated, more of the pyramidal structures were seen in the distance. From earlier reconnaissance we had overhead views which had attracted us to this location. It was one of many large conglomerations of such and so would hopefully be one of many samples.

Inside of this dune covered doorway or gateway, was a dark chamber. The chamber was immense when illuminated by our advanced LED lamps. I could see the sand had made its way in quite far. It did not offer any terrible resistance to our exploration, but it would have to be dealt with so very carefully. We surely did not want to disturb anything that could be important to our research. We entered farther and realized that it easily could take years to explore all of this if the previous images were correct in the number and sizes of structures here.

It made me ponder the first discoveries of ancient Egypt and I realized how many generations of archaeologists it had taken and was still taking to learn of that lost civilization. This too was proving to be far from any simple endeavor that one initially feels while the naive excitement prevails. So much is here, 'still under the sands' as is said even these days in the great expanses of Egypt's Sahara's dunes.

We entered in as far as we could without compromising our safety and instrumentation. Upon the walls we could see what appeared to be etchings within the stones of the interior. Intriguing, this actual petroglyph series right before us was of the giants. It was high up for those dark angels who had been of such great height and what would be their eye-level to thus render. They seemed to be almost, but not quite, writing as we would think in Sumerian or the later languages, but with a hint of the hieroglyphs of the ancient Egyptians. The pictographs flowed and undulated with an entropy enough repetition and non-repetition, so as to possibly be language. Fascinating what histories may lie within these 'writings' before us.

I remember some extrapolations of possibilities that we in different committees had often discussed, if anything substantial was discovered. Already, these initial finds were getting us heady with wanting to take in as much as possible no matter how limited our means, even with the Hypersphere.

We had come to merely explore with possibilities and now were able to at least contemplate some confirmation. But with such overwhelming validation at such an early stage, we looked toward the other pyramidal forms and any other ruins, and realized we might have to have a series of expeditions. Even with the speed of Hypergeometric Mechanics the immensity of work would take years. As recorded images had shown preferential areas of interest while others void of anything at least to the naked eye, we were able to prioritize and get more of an accurate idea of the surface area from our own observations than from the previous records.

The sand dunes here had millennia to cover and fill as the wind eroded ruins and other artifacts more delicate. Thus it left many gaps in understanding our origins, and these angels from the sky as well.

As in discovery, especially within deep time and space, we were grateful to get a glimpse enough to enlighten us. So upon such conjectures that afore had been considered myth by many, we now had some evidence.

Images and samples were collected and volumes of notes put together as we returned to the Hypersphere for our temporary abode. We would sleep and try to get used to the twenty-eight hour day and night cycle here. It might have been shorter in ages past, but the planet and its sun of a binary pair were all around one billion years older than our sun, Earth and family of planets.

There is no moon here. Alpha Centauri A is its primary sun, while B is the neighboring sun. Proxima, if one would search the sky was a bright red sun, but noticeably farther away. Over by the constellation Cassiopeia was another bright star, our own sun, and Earth where all of our civilizations' triumphs and tragedies were. We would endeavor to stay on our mission for at least a few more days, twenty-four hour Earth days. We would come back, even prepare a much larger expedition and return as often as needed. This round trip of approximately six minutes was no problem, but it was the smallness of the Hypersphere that was our major restriction. Perhaps if we could do a thorough, general survey and prioritize the more interesting spots to explore, it would be the wisest of tactical moves.

Transmitting back to Earth was interesting for if we remained upon the limits of light, it would take around 4.32 years for our signals to travel. We would arrive back before they did. It was expedient to send the transmissions through the chord of our entanglement. Thus it would be more in temporal phase of our practical kinetics in this grand venture.

Upon our retiring within the Hypersphere for the end of this day of business, I had received a transmission of an overhead view of some very interesting areas not far away. One seemed to be of ruins and two pyramidal structures all close together.

We could send a robotic craft to get a closer look as we had done here in our present location. Thus, killing two birds with one stone, so to speak, this would maintain our most optimal strategy and allow us to prepare the next wave of exploration with the most efficiency.

We slept well, the entanglement was maintained, and all the support systems that were required to handle data and to interface back and forth with colleagues functioned well. We even had acquired some small samples and many images to send back.

The next day we headed for the nearby structures that we had targeted yesterday. It was about an hour walk and when we entered the area it was wonderfully clear of dunes. We could make out clear, open gateways. So far any door-like contraptions had not been discovered, only the open entrances that allowed easy access within the enclosed pyramids. Large artifacts were soon found apparently of a size made to be used by similarly large creatures ... large beam-like poles and spiked, beveled iron pieces, some harder than iron, possibly metal alloys.

Everything seemed very huge to accommodate the greater size of what we ascertained as the more brutish of the giants. The more thin, delicate, intelligent administrators and leaders were utilizing far smaller almost human-scale implements. There seemed to be an intermediate size, that may have been overseers, we hypothesized. There were images upon these stone surfaces as well, petroglyphs we would need to interpret to understand more about these beings. There was so much to analyze in substance that it would require years to comprehend in context of all past implications in history and archaeology which we had recognized for centuries.

It is ironic that we strive to understand events from glimpses and pieces from ages past. Then suddenly with a great discovery we are overwhelmed with such an influx of new evidence. The quest for insight from the information we have been seeking is buried somewhere under so much to analyze. We either strive in a desert, or overcome in a flood   of information; also struggle in a feast or famine of understanding.

We rethought our tactics for further exploration as we looked over so many artifacts and articulated ruins before us. As we closed in on one of the entrance ways before us, the inside was dark. We would have to illuminate the interior if we wanted to see anything.

The builders had lived up to their nickname, but their similarities and differences to humans were fascinating to try to realize. Here inspecting these larger than human life implements and their equally massive buildings, we were face to face with the work of the giants or dark angels from the sky. We moved inside one of the larger but damaged pyramids. Noticing that our Geiger-Müller Counters recorded increasing radiation in the area, we became quite intrigued.

We remembered the Oko supposedly natural nuclear reactor in West Africa, known since the early twentieth century, over one hundred years ago. What was fascinating was that it seemed to be about a billion years old with strange, enlarged skulls found nearby. Was this the realm of giants, the fallen angels even then? We had understood that other similar 'natural' conglomerations of uranium were also discovered at about the same time. Now we were detecting radioactive emanations and though the levels were low, it made us wonder how strong they had been in their original time of concentration. Was this around a billion years old too? If true, then it would fit with Alpha Centauri's stars being at their prime about a billion years ahead of our sun and Earth. Perhaps part of this pre-flood time was becoming more understandable.

We had long ago contemplated that the so called 'global warming' hysteria of the late twentieth and early twenty-first centuries was not entirely based on the variability of the Sun's output. Now we were able to ascertain great time cycles than were the ones based on eighteenth century and earlier assumptions.

We had learned through some of our explorations that the Sun may be older than what was convenient to consider. If that was the case, with our higher resolution of measurements, the Sun had at times been burning some helium. This meant that along with being a sibling of the Alpha Centauri stars, they were sharing a far more similar age. Our sun may only be younger by millions, not billions of year. Being a main sequence star and much farther along in middle age, the Sun had significantly less of a temporal window to remain benevolent to our complexity of life on Earth; we have been able to presume it to be about 500 or more millions of years old.

There were many reasons we were finally in our New Time Millennium, and the New Administration of the One, Ha Atik Yomin. Now, without the profiteering of warfare and human-controlled government, we were heading to the stars. Alpha Centauri was close and nearby enough to explore. When our scientists could better calculate the coming red giant stage of our home sun, then we had to consider other younger benign stars to spread life from Earth. For it seems that it came to Earth from afar in similar circumstances.

God's created, natural world extended in deep time and space from far past to far future, with cycles that our small human experience does not actually comprehend. We can presume anything and even offer what we call correlation, which has to be verified by repeatable scientific evidence. Outside such is 'circular' argument, and religions though being denied, are still systems of belief to the beholder. With such, even the Easter Bunny can for so long and adamantly be argued by 'rationalist' willing to get the attention of defiance they strive for. Moving beyond such adolescent 'academics', one must pursue mandatory verifiable, repeatable evidence instead of referring to some untested source. Most rationalists would rather avoid that hard work especially when impossible to attain.

Presently we are headed to the stars riding on Hypergeometric Mechanics, so long ignored to be measured and observed with the required resolution. Some are considering Epsilon Eridani and Tau Ceti as possible new homes for the expansion of life. These stars at around ten light years away and single stars on the main sequence are like a far younger version of our sun.

As God, in His Infinite Wisdom, brought from the stars to Earth and our solar system, the carbon-based life as we know it; He is having us take life from our present time-space abode to another space-time to proliferate This must have occurred countless times in the deep time and deep space far beyond and that which we are allowed to glimpse, and near enough for us to substantiate.

Fascinating in all of this, we are finding evidence of the fallen, dark angels or the giants as we have so often labeled them. For greater understanding of our origins within the remote local Galactic Ecosystem of our Milky Way Galaxy, we can better glimpse this grand stage of life. Amongst the suns before our sun, and the planets before our planet Earth, and long after such is gone, the continuum persists.

As in Quantum Mechanics, when observing and measuring the very small and fast and/or the very large and slow, we find it more difficult to ascertain past from future or future from past. The present moment becomes more like a flow within eternity. Reference points required for up or down, past or future and here or there, are increasingly less concrete. In the extreme cold near Absolute Zero even atoms start to occupy the same place, superpostioning; thus here and there 'smearing'.

Such dissertations were often our discussions at the University. On this planet in the solar system of Alpha Centauri A, I could not help but reflect upon its tenets again.

We were to spend about a week before our return to Earth and our sun, and then take the needed time to properly digest all that we had learned and discovered.

Chapter 13

Comparison of Present Discoveries to Ancient Writings

It became necessary for me to delve into what some had with prejudice, called myth and legend. In Genesis chapter six and other Biblical, as well as secular works, similar characters were mentioned. They had come from the sky -- these builders, these dark, fallen angels -- and seemed to have come through space to a person of the recent past and this present century. They were intelligent, technological and industrious. According to what was written, they had been banished from God, by God ... humankind choosing to go their own way, which was illusion. Why? Because it is written: "You can choose between two masters." And humankind had chosen what their meager, limited free-will desired, but from the wrong master. Human pride caused us to feel the need for self-decision, even when deluded by the direction of its new master. We chose the wider, more convenient path, than previously provided by our Creator.

We need to be, or at least assume that we are at the helm of our own ship, a ship we never designed or built. Oh, how we so want to defy authority, especially the benevolent; for it appears weak and vulnerable, because it is kind, loving and long suffering. No, we want our own way, and in Eden we made the choice, and God gave us his forewarned consequences. Once on our own, outside of Eden, we struggled and suffered. But any one or all of us could have tried to make contact with Him, our Creator and choose again to submit and obey Him who knows us better than we know ourselves.

In the succeeding generations there were a few that returned to Him, but so many chose not to. These would rather choose the more convenient of the masters, though far more malevolent. Ultimately coming upon this Earth was one called 'HaSatan'. He was most handsome and charming, and had such a charming song to sing to so many of us who desired his alluring urgings.

To the fallen angels also banished from the Heaven of heavens, we were but another creature of God to 'domesticate'. Because of the allure, we sought to be independent. But to so retrain ourselves, we had to defy God, and many did. For the small remnant of humankind that did continue to worship God, El Shaddai, many would be shunned and even persecuted. We few that struggled in our obeisance to God would be the inspiration for a few more here and there, whilst a severe burden for the defiant.

The human race was choosing those from the sky. It was gratifying to have a shortcut to the better things of life. The bait of the fallen angels had worked well. With time, their Prince, HaSatan, would gain the upper hand. As the builders continued to construct, eventually more humans would try to be like God. But they never could, for they were reflecting their 'fallen father', Satan, and soon desired to build into the sky themselves.

The breeding and farming of humans was rampant, and for some it was without restraint, so desirous and egocentric they had become. Soon a tower to heaven was being built. Could the pyramids, the obelisks and other structures be representative of vehicles to leave the Earth? Were humans attempting, with a more sophisticated technology to prey upon another world and replicate their monstrous themes?

Shortly disaster came upon the Earth as a literal 'flood of judgment'. Afterward the ice commenced and froze globally what remained or was not buried deeply enough. Some survived and intrigued us by their rediscoveries ages later.

With God as Master, one has all the full potentials and moderations for proper conduct. He as our Teacher and Father can help and guide us in peace and love and virtue. Even so, there would be very few that returned to Him.

Over more ages, we humans attempted the power of technology without our Benevolent Master and chose to do it ourselves, but again the delusion was the result of the seduction. Thus we faced space once more in our very advanced state, but relying on our own understanding, to again follow the wrong master.

In the blink of an eye, many vanished from the Earth, but without any corrupted technology of Godless design. Rather, they were gone by His wisdom and intent far beyond anything a fallen angel or HaSatan could ever do. Before my time during the Great Age of the One was the sudden 'going up' of so many at once. We came at a very difficult period. After this, the One as Master, Teacher and Messiah returned. Presently we have been better able to glimpse the grand continuum of enough deep time and deep space so as to better understand our place of submission and fulfilled gifts and talents within each one of us.

This was leading up to what we were preparing for. The Sun was slowly fusing helium. There would be a time after this period of a Post-Edenic Time, that we would in His will propagate into the heavens and beyond what we could imagine. In this Post-Millenium Time, the Sun would expand across the sky as a bloated, red giant heating the Earth to thousands of degrees, boiling its oceans and eroding its atmosphere by the fiery orb's more powerful solar winds.

For a time Europa's oceans would warm and have surface currents and waves. The atmosphere would increase and the new epoch of our sun would be a welcoming influence.

Then as time passed, the Earth would feel the impact of its moon coursing so far away in its orbit, swinging back and colliding with the dry, hot, rocky, Venus-like Earth. This would persist for a time, but the increasing Sun would slowly consume within its bloated atmosphere, the shattered, tectonic, latter Earth. Here all major human remains and ruins would be reconstituted within the heat of our very aged sun and all would be as if it had not been to the human perspective.

To St. Augustine's 'God is the maker of worlds', I would like to add, 'and the destroyer of such'.

From approximately 120,000 BC to 120,000 AD would be the designated rise and fall of humankind ... all of our arrogance but a speck in the deep time and space of God's eternity. This was the consensus of us all, but with details to come that differed only slightly because of the limits of speculation.

Some found it intriguing that Mercury was the core of an ancient, solar 'hot Neptune' or some large gas or giant ice planet. This ascertains an older sun; that when the planet slowly made its way from the colder parts of our infant solar system, the Sun had an accordingly, effected wobble to its axis. But over great lengths of time, the larger planet in its orbit lost most of its gases, due to the stronger solar wind, its closer range of elliptical orbit and the Sun's wobble, reducing it to what is in line with most of the planets near parallel lines of orbital axis. Thus our far older sun matches well with the open cluster of its siblings, the Alpha Centauri stars.

The voyage of 4.23 light years was reduced by Hypergeometric Mechanics to just nearly six minutes. This fantastic traverse had brought us face to face with the ruins imaged amongst other objects upon the terrestrial-sized planet orbiting Alpha Centauri A and the one orbiting Alpha Centauri B. The one at A was a bit larger and at 3/4 the Earth's size had the most Earth-like conditions and the most ruins; while the one at B was smaller and similar in size and description of our Mars.

So within this solar system was the historic origin of the giants, with two stars nearby and a third much farther away. Proxima Centauri, with other terrestrial planets, had quite a volume in which to proliferate. I say historic because of their connection with very ancient Earth history. To better offer context, the histories of the Sun and Alpha Centauri interfaced and were probably only samples of such at greater interstellar scales. Where the giants came from to Alpha Centauri was another mystery; God only knows.

These dark angels or giants learned well their celestial mechanics and how to inhabit different planetary environments around three different types of stars. Proxima Centauri even had gas giant planets and offered more moons and Mercury-like terrestrial planets with one in tidal lock. Around both Alpha Centauri A and B, the retinue of planets were terrestrial and similar to our moon and Mercury. There were only a few planets orbiting each of these two stars.

Nearer to A and B there was another outer, Neptune-like planet seemingly detected, maybe two, at quite a distance. Siblings of the stars were somewhat different but similar. Just like human siblings, both had some generally shared characteristics and yet each had its unique qualities that made it special in its own way.

On the greater scales we tried to view, we realized that we were probabilistically sampling in the same proportions of qualities. In other words, we were seeing more of ourselves and our origins in a bigger system of other life forms and their origins too. This includes the consideration of non-human intelligent and technological, as well as the whole scheme of size and complexity according to the level of benevolence of environments.

With the advent of Hypergeometric Mechanics, thank God we could now on grander scales of time and space better show the repeatable and verifiable evidence needed to establish scientific theories to becoming established scientific facts. No longer did we have to suffer the constant, circular debates of rationalists, these 'enlightened academics'. They feed to their minions 'not to be questioned facts' in order to intimidate by their presumptive avoidance of the testing of their opinions.

Many of the rationalists are limited by their own 'irrational' limits of objective reasoning. They often say that the Bible, especially the Old Testament is irrelevant, whilst not proving the relevance of Shakespeare, Socrates or any of their other pet interests. In the main, relevance is in the eye of the beholder. Irrational, circular debate is itself an anathema to a supposed rationalist. Why they senselessly pursue unprofitable winning of their arguments only attests to their obsessive-compulsive need for such. So they are finding their fulfillment, not in greater understanding or knowledge, rather by getting the attention that their inner spoilt child tyrannically demands. Try scientifically debating such an enlightened one and you will soon find their preferred wrath and your debasement, whilst their premise of argument is never attained.

They love to argue for irrational reason, and as any derelict of pseudo-intellectualism need to be given a piece of candy and told to finally defend their insanity before real scientific peers. They will find some excuse to avoid this and so, continue to suffer, only to find out that they were exposed at the onset to not be the teacher or academic marvel they assumed themselves to be. They are the last to realize they were found wanting of any intellectual worth and as annoying as a mosquito on a hot summer day.

Well, those intervals of such waste of time have been superseded by our New Time. Finally with the ability to really share repeatable verifiable evidence, there is no longer room for unproductive debate. Now we can actually portion and process the great things previously diverted from the true scientist because of biases of bullies of thought and belief. With such we can actually venture to and enjoy new worlds, and consider productively their benefit to humankind. The greater scale of things has as much to reveal as the refinement of the small scale of things.

With all that behind us, we can better see how consensus is obtained of this far past time to our time, and to propose by such proper extrapolation our future time. Thus a time- space spectrum of function and the structure upon which the function proceeds, is a continuum.

H. G. Wells' books are still available and they still so quicken the mind of thoughts. With Hypergeometric Mechanics, his considerations of time travel were possible, though it was also concurrent with spatial travel. One affects the other; they are integrated.

I have often stated, "The implications of the application of Hypergeometry to practical mechanism are provocative!"

Chapter 14

Pre-Cambrian Earth and Its Implications

With the time to absorb what impressive, yet still rather small in volume, new data we have gained from the Alpha-Centauri System and just one Earth-like planet there, we choose to let our imaginations soar. Not completely without restraint though, we lofted into the clouds of the possible. Considering our Pre-Cambrian Earth and its history from this period, we have expanded our small data set to a greater scale of time and space. We are but a part of something much grander, yet enough to glean a glimpse of a greater, amazing, complete picture.

So many of the planets we have and perhaps shall ever explore, are in some part of a Pre-Cambrian state, if we predominate in the numbers of the probability of such.

I might find one particularly atypical daisy in a large field of them, but even though each has some uniqueness, all will have the same general trend-identifying characteristics of a daisy. So let's also speculate with larger scales of space-time. We are then but an example, albeit a very special one, for all are unique, though distinctive in proportion to trend.

Most likely we will thus find hot and acidic oceans with archaeon life and its more fundamental forms predominating. Now with the scientific basis of Inherent Differentiation, replacing the old data-starved 'evolutionary theory', with its better foundation in mathematics, we can better see what life has derived from. And God is still in the details and life's Originator!

How? Well, in the early twenty-first century, astro-biologist had discovered the essentials of paleo-biochemistry. Here experiments had been performed with proteins and it was found they have a consensus for certain amino acids. With such we can use large numbers for trends. We can also parallel these functions with Post-selection mathematical functions used in solving quantum erasure and relativistic problems that seem counterintuitive, such as traveling back in time and killing one's grandfather.

With the Post-selection function only the true state is allowed to be the answer, for all the others have to be false. What this does then is eliminate the impossible from the possible. Back to life as a non-entropic function itself, then we also see where a forward time, entropic, and reverse time, non-entropic progression end up as the same progression. So if you exist and can travel back in time, you cannot kill your grandfather, for your mathematical function would falsify itself. Thus in Hypergeometric Mechanics we now have world lines of alternate histories existing per function, with each within itself, true in forward and reverse. Therefore, $t \rightarrow -t$, the time forward and reverse equivalency formula.

With Inherent Differentiation being triggered by only two stimuli, environment and time, with subsets only within these two domains, life is a topological function that unfolds in adaption to its time and space of existence. This is also saying that all that is in its DNA/RNA is already there to unfold because life's genetic material is a hardware/software computer, programmed by its Author and Creator-God, throughout His Universe and Superuniverse!

Considering this within our Milky Way's Galactic-ecosystem of time and space many billions of years before and after humankind, then we see His great work as a fly looking from above upon the Mona Lisa painting. While the fly was on the painting all the globs of color and topography did not make sense. This fly, yes, a very intelligent fly, now sees from another dimension of its spatial geometric the whole, bigger picture. This context now makes sense of what on the local smaller scale seemed so counterintuitive.

With our use of the Omega-Aleph ($\Omega\aleph$) constant for superposition of multiple states in Hypergeometric Mechanics (which replaces zero, infinity, $\sqrt{-1}$ and 1) we have no room for randomness. Thus determinism is established despite its consequence of limiting free-will and allowing one to ignore micro and macro states beyond one's ability to have the resolution to measure.

All is mathematics, and also a symphony of symphonies. The One who designed this 'music box' also built it and maintains it.

All this is similar to finding the linguistic root of a modern language, like French or Spanish, which root to Latin, and all of the ancient languages root to a more ancient tongue origin. As Ezekiel's mention of 'wheels within wheels', we understand in far more ways than first impression would allow.

This consensus-seeking of only certain amino acids by proteins is but a shadow of deeper functions mathematically.

We are in a very important place of certain time and space parameters within a far grander continuum, for the smallest to the biggest, the fastest to the slowest and the nearest to the farthest. This significant place is important to us for sure locally, but also to all within a universal sense. In human perspectives, we often equate the enlarged volume of time and space more important and smaller time and space to lesser significance. We are so subtly biased that to be human and totally objective is impossible. Thus, more of the bane of the rationalist philosophy, for they are in containment of subjective and irrational biases, while in denial of such; yet it is written: "wisdom is justified by her children." The self-assessment, supposedly peer- review of such a cluster, has only to be allowed fruition. Then we see that despite their obstinacy, they are of their blindness and deafness so proved.

I so long for the progress to the stars, whilst at the same time, I would starve and stagnate in neo-modern, humanistic- stifled, human philosophies. Even though those times of ignorance are past, its tyrannies of mind are archived for us to analyze.

In the far reaches of a universe we call the superuniverse we could at this moment glimpse more. And the astounding implications were such that even the temporal dimension, possibly even more than one.

It had come up in the early twenty-first century of a possible hypothesis for a second dimension of time. This at ninety degrees to the time we seem to better understand, would supposedly be of the time dimension similar to the other four dimensions. What may be inherently different is the respective fifth dimension of Inertial Geometry, once called 'gravity'. The second dimension of time may have three other spatial dimensions and a fifth dimension of Inertial Geometry, but to us it may be of the appearance of anti-gravity or anti-Inertial Geometry. This could appear to us as some of the expansion of the superuniverse and seen in a minute amount of the Red Shift, local and observable universe upon this mantle of non-inertial rotation. To us this mantle is the Tachyon World. To the other side of our superuniverse's Event Horizon, we seem a Tachyon World as well. For each of our respective sides is within the mantle on the other side of our respective Event Horizon surfaces. That surface of course equals c, the speed of light.

Back to this possible second dimension of time, which could also be the Absolute Time that Sir Isaac Newton so admired. It could possibly be a reference of sorts in these very grand extrapolations that would require further resolution of research. Using Hyperspheres in more complex and extensive explorations could possibly reveal additional relevance, or irrelevance, of all of this.

What we have more substantially though is the relevance of Hypergeometric Mechanics and our ability to traverse great distances in reverse time at 370,000 times the speed of light. It is possible within our present scale of time-space exploration to easily travel for distances of hundreds and thousands of light years in functional scales within a human lifetime. This is where we are presently, yet with our further glimpse through the proverbial dark glass, it hints at even more.

With this far larger scale to view, let us try to focus in on some more of the possibilities humanity would be able to handle. We shall ponder my churning reflections upon the proliferation of Pre-

Cambrian states of age and development of other Earth-like planets, and the flora and fauna therein, probabilistically.

We have already done some interstellar expeditions to the Alpha Centauri System, but there are others. Tau Ceti, and Epsilon Eridani, and Epsilon Indi and more, but these are farther away. We have expanded our search to F and M stars, the orange and red ones respectively. F stars being hotter are rarer than the G type like our sun, and then there are the many M dwarf stars. It seems that proportionally the brighter, higher energy stars tend to be fewer than the dimmer lower energy stars, and accordingly their masses characterize the same. So let us try to limit our considerations to carbon based life, which is the more adaptable of any other derivations, like silicon, though all have some expression within great scales of time-space, I am sure. Considering just the F, G and M stars, we can see that in our Milky Way Galaxy, and thus other similar, spiral galaxies, God's natural world is well diminished from restrictions.

These other galactic eco-systems, like our own Milky Way, are of fantastic diversity, having similar lines of possibility as our own Earth. We have islands in the tropics and also in the Polar Regions. We have continents, ice sheets, deserts and oceans, shallow salt seas and fresh water lakes ... so many theaters of similarities. Even in intelligence and tool making there is a human (made only in God's image) element; whilst other lesser forms such as apes, otters, ants and sea life that are able to articulate their environment with extra-biological amendments to survive and proliferate. Therefore tools can be considered or redefined as biological extensions.

So we are not shocked at any non-human forms of intelligence or of tool making, and together they don't send us to any lack of consolation. We rather find such exhibitions natural and even entertaining in some cases. Familiarity breeds comfort, so if we may dare to give God's latitude from our restraints of ignorance, then other worlds with non-human intelligence and technology should not be too troublesome.

Often, I had wondered how little we dare to think and still believe in God and so easily rely too much on our own limited understandings in our pasts' history. No wonder, that when we were driven out of Eden, our own arrogant defiance already was starting to limit our potentials and actually become dangerous for the rest of Eden.

Earth may have been somewhat more massive originally. Conceivably another smaller planet could have collided with our larger planet, making our moon out of the coalescing debris strewn in orbit. Some of it could have been ejected from Earth's gravity, its time-space curvature depression. Possibly then much more of Earth may have been required initially. This may mean that our Earth was possibly two or even three times its present mass we conjecture at this time.

Having said this now and in the book, *Hypershere ... A Journey at the Speed of Geometry*, it was implied in the twenty-first century. The author maintained that more likely at two times the present mass. The reason for this was that in a three times the present mass scenario, the atmosphere would be far thicker and could possibly hinder photosynthesis of plants; and with a stronger Inertial Geometric, then life would be hindered in its mega fauna extrapolations.

Then let's consider along with the afore-mentioned author and his book, that a two times model fits much better. In both models, hydrogen and helium would begin to be held in the atmosphere. At some point, maybe around 2½ times a slight mini-Neptune effect for this somewhat larger Earth. Perhaps with more sunlight able to filter down, extra water in the atmosphere, more gravity and a stronger magnetic field, it might have been preclusive to a slightly more benign environment. More oxygen could possibly help too, with the metabolisms needing greater amounts under these conditions.

This Pre-Cambrian Earth (and similarly for any Earth-like planets in that time frame, whether smaller by a fourth to a half of our present Earth's mass, or up to two or three times its mass) would probably be the most timely find of Earth-like planets in God's natural world. For that matter in our entire Solar System, in which are also other suns amongst our Galaxy. This would then seem the trend of such geometries to other spiral galaxies.

Now if we reverse our perspective somewhat, then the giants, dark angels or builders from the sky, would seem to have arrived during the Pre-Cambrian time on our Earth's history. And the circumstantial-like evidence would be the Oko natural nuclear reactors in Africa and the large skulls found in the region of the giants. These all seem to indicate activity anomalies for their time, which ascertain as the Pre-Cambrian, for they are in two to three billion years in age. Thus, even the Earth's nearly 7/8 of history is in the Pre-Cambrian, which makes for very significant interpretations.

This would be the greatly probabilistic precedent that most exoplanets would be found to be in their Pre-Cambrian-like state. Complexity seems to be proportional to age and rare to the quantity of the numbers. So let's try to keep in that venue of considerations. So far, that is what we have even with our Hypergeometric advantages in propulsive technologies.

This was and had been one of the many dissertations in academia that we would hold. This time it was after our venture to the Alpha Centauri System. Mind stretching exercises like this were great. To extend any additional piece to the puzzle of the human story was always personal and wondrous for all of us.

The liking for the geometric chord of a circumference is the best way to reckon 370,000 times the speed of light (c). I shall delve into that with the aforementioned circle in a moment. If we look unto a distant point in the universe, say to Alpha Centauri from our Earth, and knowing its light distance, it gives us a partial length of a circumference. This light distance is approximately 4.32 light years. Now the chord is the distance between two points, and in this case it is around six light minutes, at 370,000 times the speed of light. The relationship here is that the chord is about 1/370,000 to the light distance. This ratio is universal as far as we know from a circumference equaling any light distance as a denominator to a numerator equaling the chord distance. That is quite impressive! So inverting the chord is at 370,000 times the speed of light nearly *2.7027027... x 10 e -6* the distance of the light distance. This shorter distance is also in reverse time, as I have repeated in all these dissertations within this grander story.

For a moment, while we can see the Alpha Centauri Stellar System, the light follows the longer surface and the chord would appear a long curved link imagined off to one side. But in reality the longer line to Alpha Centauri, though appearing straight, is really the curved geometric partial circumference and the chord is the straighter and shorter of the two. This is invisible, but to the eye of the imagination, only a concept.

So counterintuitive and that is why it has taken humankind so long, except for the grace of God, to figure this out. Perhaps what we experience as inertia is thusly different per immediate time and space curvature on this grand Event Horizon in other locales.

This commonality of the ratio of the partial circumference being 370,000 times longer and curved to the chord, gives us some insight into the mantle of the Tachyon World underneath, where this Hypergeometric Mechanics takes place. In the Tachyon World one can dare to call it some standard 'state of rest', at 370,000 times the speed of light and in reverse time as its normality. Without this Tachyon substrate, our four dimensional, local and observable universe would not exist, at least as we know it.

There is a distortion in all this and so our circle with the remaining circumference is a bit squashed. We may be sensing a spinning of the superuniverse where we are located near the equator of this spinning black hole. We have to speculate here for the intrigues come from contemplating the associating simple geometries that seem to come to mind ... the Inertial Geometric and the resulting inertial aspects.

Oddly enough the inversion result of *2.7027027027 ... x 110 e − 6* previously mentioned, relates to the Curie (Ci) of radiation decay. Perhaps, some of the source of the background radiation we can measure?

On this squashed circumference, perhaps very spherical, we are obliged to accept that we are measuring the greater remnant of the circumference. If it is a constant, then is it the diameter, the longest chord? We are still experimenting with all these concepts. One glimpse or test just leads to another and all more extrapolated from the least we may be able to see.

Back to our overview of the likely predominate Pre-Cambrian states of other Earth-like planets. We seem to be at a fairly good distance from our Galactic core, somewhere around 27,000 light years ago; this means we may be a bit younger than some of the stars nearer in radius to our Galaxy's nucleus. Earth may be older than many stars, all of this statistically supposed from our Galactic orbit, but farther out in radius. This may mean probabilistically that we have along our Galactic orbit the most similar in age and state of other stars and any of their Earth-like planets. Most likely any sibling of our sun from her open cluster birthplace, would also seem to orbit most near this radius.

These factors have over time made the stars of Alpha Centauri very suspect as siblings, along with age and astro-chemistry proportions.

Reflections upon reflections ... we had our summits of academics for weeks with schedules for such and to then do our work in our respective routines. So I took some more time to relax my thinking some; though as a thinker, that was easier said than done. Finally with some quiet time I could let slowly what was digested surface at a comfortable rate and let my mind wander on here and now things. I could go out to dinner or watch some sport and entertainment. While indulging in such, slowly, as is my mental characteristic, more completions in things would often come to mind.

Neat stuff, this mind of man or as we often said in these times – 'it is a quantum/relativistic machine and the body a vehicle'. In this clearer time of understanding, we could say, as King David had said, "How wonderfully am I made."

Chapter 15

Future Apocalypse and Post-Cambrian Earth

The Sun has a lifetime. Her planets particularly our Earth, one of her retinue, have a lifetime as well. The Sun at some point shall burn helium and bloat in her size and heat, dissolving the inner terrestrial planets, including with high probability Earth. Mars well may survive in an oven-like stage.

The outer planets may suffer severe solar winds that possibly will erode Jupiter's and some of Saturn's atmospheres to result in lesser diameters, making them quite interesting as larger, inner planets. Uranus and Neptune will warm; feasibly Pluto and the other ice dwarfs may warm some, with lots of water to melt from the primordial ices. I think of Europa as with some atmosphere and a liquid surface. Perhaps not for long term, but enough to be a fascinating subject for any others of the non-human technologies class that may also be exploring from the other nearby stars of the time.

Absolutely, I am not quite being tongue in cheek by calling this time the Post-Cambrian, for it may well be the future mirror in a reverse process of what the Pre-Cambrian was, and may be for almost as long. And from afar, the two may be quite indistinguishable until one makes more resolution in the surface conditions of the planets of the time. There should be some remnants of past geologic activity, fossils and atmospheric isotope signatures that trace back to earlier times. This reverse mirror is not identical in exact replication of detail, but of trend more so. Very interestingly, Poincaré had in his Recurrence Theorem, the figure of *10x e120* for the whole universe. But that would be the longest considered, yet the recurrence would need such to more exact in its replication.

I had once seen in old books, someone had said that Poincaré may have needed only *10x e32* for enough of a replication of history in the universe and for us to realize alternate world lines or alternate histories. What many may call a 'multi-verse' is really these alternate world lines. Add to that if you will, these isolated local universes upon this superuniverse, all rotating as a whole and being actually Hypergeometric in construct and function, is the reality we are experiencing. No the more meager local time and space geometry we experience every day, has been more understandably projected by this grander 5-D form. All of this so repeated in other writings and in this writing, now seems to make much more sense. All so mathematically based, and needed for what we would understand as the foundation of space-time itself.

These theories far more ensure determinism mathematically and thus call for the reconsideration by many of a randomness that allows them to excuse their responsibilities and to allow unprovable thesis to their philosophies of life. It is so easy without repeatable scientific evidence to accept anything for one's own palate of taste, yet when based on mathematics, measurement and related observation, can better be understood in truer form.

Thinking along a horizontal line of time symmetry, $t \rightarrow -t$ allows one to imagine a line either past to future, or realigned to future to past. While imbedded in one line of flow or the other, the unimpeded line allows duration in reverse time, while the line one is inhabiting appears normal and forward flowing.

So that is what Hypergeometric Mechanics allows when pursued. This temporal and spatial understanding permitted humanity to finally achieve greater things, with the guidance from the 'One Waited For'.

All this, within a better world where we are as we were meant to be, has been also a time of peace and growth. Science and art has flourished; moral principles thrive to the point that justice is everywhere; no matter how local and personal, the influence has spread to the full extent of our world.

We had learned as it is written: "There is a great cloud of witnesses", in time and space. All occurrences in time are recorded in space and space is recorded in time, whilst all is undergirded by geometry, especially Hypergeometry of time-space and space-time. Without geometry, there is no space-time, and without God there is nothing. With God there is everything.

Humankind throughout history has shown its limited potentials with self-government, cellular phone systems, automobiles and casinos. In defiance of their Creator-God, humanity became what it was … in degeneration.

What of the future of humankind? The next 120,000 years should tell. We now look forward to and try to understand what is to come. We here in this Golden Millennium of Enlightenment and Truth, are to move on above the sky. Somewhere between Earth and its atmosphere, is the first heaven and within the space-time often called just space. The second heaven, just 'under' the third heaven that some call the 'Abode of God', will be the 'New City'. Now as this progresses farther into the future, which we can only now glimpse, we look upon the ones remaining for the continued history of the Earth unto the time of 120,000 AD.

This would probably be the best we could extrapolate of human potential and failing that would make sense to one of the time, now for us, upon such reflections.

The Sun shall surely course another approximately 120 light years. Thus given the average of one thousand years to traverse a light year in our sun's orbit around our Galaxy at our Galactic radius of about 27,000 light years from the nucleus, further events will continue to perpetuate. Here on Earth the climate will cool and heat as all the cycles continue, but the slowing of Earth's rotation over time, will be barely noticeable without fine measurement. Seasons will continue on their course along with the climate, as well as migrations of the birds, insects and beasts.

Consider humanity, no longer under the influence of a God who has taken His Own from the planet, still under the consequences of His laws of nature alone. Even the creatures who had not needed to 'learn' of God, for they always heard His voice, would be in far better condition than descendants of those who defied their once personal Creator.

The generations-replicated human would become such a deteriorated derivative, that we would possibly have difficulty trying to relate to such future ones.

We have always looked to the future for progress and forget that entropy, the Second Law of Thermodynamics, wins in any system not articulated properly against it. So to, future humanity, now abandoned of God as its ancestors so desired, continues to digress to its own species' death.

In 120,000 AD, which is about three to six thousand generations, the decadence of not only moral and ethical fiber, but of the well-being of physical, emotional and mental health will be compromised. This failed species, becomes more the fodder of those other creatures upon the Earth with whom humanity shares.

The Sun would have warmed slightly and the Moon expanded somewhat from its present distance from the Earth. Even the Earth would have moved slightly away from the Sun over time as it lost mass, and the resulting celestial mechanics followed their function.

In time with the Sun's increasing diameter and reddening, life within the Solar System would vastly change. The outer planets and their moons would be the new 'Goldilocks Zone'. From this present time of writing to another 500 million years or so, things would be generally the same, with minute to more perceptible changes for life on this Earth and the inhabitants therein.

In the course of the scale of the next 120,000 years, humankind's descendants shall continue, as those of our Special Time are in less than these 1,000 years all gone. We shall have obtained our greatest of potentials in our inherent design, and onward to the stars and beyond. Already we are questing of such. As myth and legend are the seeds of realities of long ago, thus the same to the realities of time to come. For we would return to Eden and its completion of fulfillment, now without the ancient of distractions to seduce us off our proper path.

In this Millennium of our time of greater access to Him and to our eternal inheritance of program, we have the realization of the things to come that those before heard about. They in their time saw through the glass darkly, but now we see more clearly, and much more is to come, which we still see through that same glass darkly.

Perhaps upon other worlds and within their proper times, are others as well as us, having had to learn over eons of how their position is tested to its appropriate place to the Creator of all.

There is a better understanding that the Local Inertial Geometry and its kinetic signal is required to be strong enough above their general background interfacing 'noise' of the Local universe and superuniverse, so that the present and 'now' are what is experienced primarily. This was very difficult even for the discoverer of Hypergeometric Mechanics initially. At first his own neural biases made it appear like the future was stronger than the past. Later, with even greater resolution, finally he could see the reality more clearly. The glass sometimes is very dark that we need to peer through.

The present is thus more pronounced than the past, from whereas the information to arrive in the future is less than the proper proportion to the distance. Likewise then, to the future where the information arrives; but to return from the future to the past is the weakest of the course of such information in reverse, with respect to the states of such Block Time to the other. It appears that future arrives and seems to catch the mind and its senses first as we are programed more for what is to come; but that illusion is in order to survive all similar conditions and creatures we cohabit with in the present.

In this grand design, whilst the removal of the biases of the neural-based perceptive functions, the past is stronger than the future. In the reality of resolute measurement, it especially needs to be so close to Absolute Zero. Only its processing of the brain makes it appear to be going away, as the future appears to be arriving. It is important for the cat to catch the mouse, and both with the same biases neurally. Along with the entropic flow of time, forward survival for one, the other to be eaten. For all are built to function neurologically the same, thus utilizing the inertial/kinetic signal strength at its most advantage for living systems. So wondrously are we made, as King David eluded.

By the 'End of Times', Israel was the most advanced technological nation on Earth. By 2010 AD it already had all the information of Hypergeometric Mechanics. The rest of the world tried to catch up, but were never quite able to take hold of all the potentials that had actually been available all along throughout human history. The other nations were so impressive and much larger, whilst Israel was cornered and intimidated constantly, often by even her few allies.

We of this Golden Time were the few who by design had to previously hide because our lives and our abilities were at limits within the generations before. We were the ones that many would trounce upon, oppress, ridicule and even destroy unto martyrdom, because we were so different.

It is written: "The meek shall inherit the Earth", and it has come to pass. Being meek was within each of us -- really as steel covered in velvet. We of those many generations had to learn to withstand, despite being left in the dust of the cruelness and jealousy of others of our own kind. Then in the 'whisper' of the whispers, in the 'silence' within the silences, within the 'shadow' between the shadows, we could hear and see that which others could not. It was their choice, while we were forced by those stronger than us and yet still guided by God, had nothing left but to see and hear such.

A generation was beckoned to build these things, before some of these shapes of the things to come, which when they arrived were stolen away, and all about them destroyed. It took a child-like attitude, and in many cases just children, who were able to understand these greater things and even experience them; while the large scientific institutions were stumbling over their stubborn assumptions and more importantly their political and corporate careers, agendas and profits.

One is so easily blinded by presumptions, especially sophisticated ones. It is written: "lean not to your own understandings", rather one should test and debate until there is repeatable scientific evidence from which to sincerely base conclusion upon. The more we glimpse into the night sky, the atomic world, the mathematics of form and function, the more our horizons expand. There are places we may visit and other places we may only contrive. Humanity has a window of scale from which it has its most efficient envelope of operation, and even bliss. But yet, our own arrogance often has limited us to such a truncation, that we choose to impose willfully.

As if again a child, an adult once more can create great works of art and literature, and openly observe and measure the science of the world and 'worlds' around them. This child within is the home of the imbedded flame that so many others try to extinguish. There are those so academically learned, whilst so practically ignorant of much that is possible in real life. Some of the greatest thoughts of geniuses who are either so ignored or condemned, that what fruition was just within their reach, are rather lost or destroyed. We could have been to the stars and beyond, long ago. Humanity casually claims originality of its inventions, while only by trial and error has it copied with embellishment what the God of Origin has made in life and inanimate function.

The return to Eden from Eden-lost has been a wonderful, severe, tragic and beautiful epic, where each of us has been destined to make a choice. Singularly in the end each has made our choice, thus determining our fullness of fruition. It is written that we 'reap what we sow', and only a few of us have chosen wisely, despite the tyranny of those who did not.

Such observations are from the ancient suns of earliest skies, to the Sun and our sky, of our past to our own future days, to future suns of future skies all so far away. Time and space, space and time are so deeply related and more intimately than ever before perceived.

Now we know, the speed of a thousand years to a day, and a day to a thousand years – 370,000 times the speed of light!

Fin 1 of 2

Midlogue

Currently we have left the expanses of 240,000 years which itself is in a grander session of time, in the millions and billions of years more impressive farther than before and farther than after. Our point of reference is our accustomed present or 'now' in its 120,000 BC to 120,000 AD midst. Here we have telescoped to early humanity, with a later telescoped point of view into the future of humanity, Pre-Edenic to Post-Edenic in perspective. Where we came from to where we are going, the past, the present and the future is as a continuum that we seem to glimpse, but take for granted.

We have so easily conceived of the scales of space. We see our small life within the context of our community which is dwarfed by some larger political-geographic entity, then our nation within some part of this planet and solar system amongst the stars of our Galaxy.

We enter now into another setting that involves a reversal of temporal traverse in part, and also a much shorter interlude. It is from 2010 AD to 1895 AD, quite counterintuitive, but fascinating at the same time. It is much more personal in effect because of what could happen if local comfortable space-time is so curved as to render one's state of causality to be quite different. All the more to ponder as the possibilities become even more personal to the one who is actually doing the voyaging. Herein we consider the more current period of our lives for exploration with time as the predominate journey. Here we honor the time of the first publication of H G Wells' book, The Time Machine in 1895 AD. Presently there are just a few sharing and knowing of this work amid the beginning period of advancing technology, as we know it.

PART II

2010 AD To 1895 AD ...
The Personal Temporal Effects of Hypergeometric Mechanics

Chapter 1

Alycia and the Golden Cities in the Sky

The pendulum is the simplest machine, not the wheel. Consider for a moment a little girl on a swing. Perhaps rather than she being held to the Earth by 'gravity', it is more accurate to say by Inertial Geometry. She is upon those moments of experience ... living on a pendulum.

To the practical sense of her parents, she is often a dreamer, but this girl already knows that only dreamers can truly 'see'. She is always late because her head is in the clouds. On her swing she is watching as the sun sets with its yellow rays painting the lower clouds in shining gold.

Only a child, not a philosopher steeped in theories of the greater substrate of the geometrics of space-time ... she just enjoys that which she has yet to further envision. Getting wonderfully lost yet in more familiar places, is surely something challenging to share yet so easily desired.

Event Horizons are in the conversations of physics, especially in the extremes of conditions. Yet, with the border between Inertial and Non-inertial Geometry ever closer to our everyday experiences, perhaps only a child can appreciate rather than perplex over such things. Wonder as a child is unlimited by the restraints of supposedly more mature pursuits of contemplating such very complex abstractions.

This had been well ascertained in 2013 AD by another 'child' who wondered if the norm in the human social context of history was of the Fourth Law of Thermodynamics. Yes, the 'Fourth' Law, by none other than me, another dreamer.

Why if the Third Law of Thermodynamics considers Absolute Zero, but more accurately, 'Absolute Non-entropy', then the Fourth Law ponders the particulars of Absolute Heat, better, Absolute Entropy. Such are the contemplations of the Arrow of Time. All this our dear little girl, Alycia, as most children while in their imaginings, has no such thoughts.

It is so grand to just swing and get lost, so to speak, in the moments of escape from the little personal tyrannies surrounding her dream, escapes and just plain unbounded childhood. What a wonder to behold that which is either ignored or invisible to others. We then can ponder and appreciate that one is no longer as Shakespeare's 'to be or not to be', but rather 'to be or not'. Now that is very profound.

So often when we children, whether adult or actual adolescent are so 'off' or 'out of it', we do not want to return too quickly.

"Get in this house, young lady," her mother scolded as she was constantly trying to keep her daughter's feet on the ground. "You have a supper that is getting cold."

"OK, OK, I am coming," cried Alycia as she rushes into the house.

Once off the swing her thoughts returned so oft to the practicalities of the present in her time and place. In the sky was how she was interpreting her escapes from the difficulties of being just a child, even with so much to be often misunderstood and underestimated. It was the chronic struggle of the unique child, the dreamer, but so very personal.

As she settled again to supper and mundane routines of the end of the day, her mind would wander to those subtle moments at the end of her swing, which for those around her was but a moment, while for her hours and even a day away.

The Sun so wonderful at the end of a day, in the morning created more golden clouds by the light of the low, bright star. She was just awakening from sleep and not able to be on her swing in the backyard just yet. But her thoughts took her there.

Those dazzling, golden cities in the sky, those people, even Grandma and Grandpa, now gone for so long ago, yet there smiling and chatting. As she visited with them, there was always abundant, other-worldly music around.

"Alycia, get your feet on the ground! Are you listening to me?" suddenly intruded her mother's tone. And as abruptly the pleasant contentment shattered. Such was the sorrow of seeing so well what others cannot.

It was the pendulum action of the swing, going from a potential energy state of a yet immeasurable moment, through the kinetic energetic swiftly to almost a stop in the next potential energy state. That our dreaming child floats away so fast, is not in the here and now at all. It is so very quick to the average eye as not to be noticed, but for the one oscillating it is more real than many would dare to ponder.

The ephemeral Event Horizons, those shadowy, evasive to measure edges of existence, are unknown to those who never wondered. No longer just in the extremes of black hole physics, but even closer than the tip of one's nose, are where some of the strangest of possibilities begin.

Perhaps as David Deutsch at Cambridge and others contemplate the multiverse, here we postulate the superuniverse of many world lines. These lines of the direction of time and all possibilities cohabit only microscopically apart from the other, yet on this side of the Great Event Horizon, all in the same direction of time. Our brains, such quantum machines of decisions, are able to choose on which line to perceive. Yet all the possibilities are of free-will and limited in range of decision. It is a mathematical flowing river still in its less resolute tendency going from one purposed point of a great history to its next.

Its destined course on the grander scale still maintains its exact mathematical functions, deterministically; a simplicity of free-will mechanics within a larger determined functioning mechanics, as with the developed Hypersphere's immense space-time consequence.

The culmination of twelve years of theoretical investigations, seven years of tedious experimentation and most recently three years of provocative results, was by 2010 AD that such was discovered to be more possible. It was as simple as a child's swing, but with frequencies of oscillations and use of electromagnetic articulations, as natural as the flow from present to past on this side of the great Event Horizon of our superuniverse.

The mirror of Inertial Geometry upon Non-inertial Geometry was porous for even a little girl. If her body was flitting from the present local to the future past Non-local, beyond most resolution for measure, yet mentally, the quantum and relativistic perceptions made her soar. Without human witness, the 'Golden Cities' in the sky were attained at 370,000 times the speed of light. She did not understand all the details, but rather enjoyed the journeys magnificently. She had become a stranger in her homeland. Dreamers never fit, they were not meant to.

Other situations could also imply the subtler aspects of our superuniverse, while shadows and whispers may be just attainable in resolution of observation and measurement. Back and forth we oscillate into and out of our present. Subsequent accounts will help to more elucidate such grander things, without which our present understanding of reality would not be possible. Such we will continue to conjecture and illustrate.

Young's Double Slit Experiments, the extremes of quantum and relativistic realms, offer us so much intrigue and amazement. We like children in our wonderings, followed by more abstract analysis of thought, open the doors to extrapolate upon the grander scales.

These things scientists are analyzing, yet like 'Alice in Wonderland', Alycia was actually experiencing.

It is the child-like wonder again of her perspective that allowed us to join her reactions to such things. To be able to just share with her mother and others, especially adults, tales of the sounds and sights seemingly near, yet so far. If they would have taken the time to look into the morning or evening sky, then they would have seen the golden shine upon the clouds as the sun touched low to the horizon. The 'senseless preoccupation' of the busy ones missing the reflective swoon of the child, Alycia, upon her swing, contrasts the focus of the beholder.

The music and paintings not by known or accredited by our supposed authorities and scholars of our 'cultured' civilization, astounded her so. The constructs were so dream-like and yet so real. The people so personable and intimate, some having been missing for generations, occupied her lonely abandonment. She was more thoughtful, sensitive and reflective, yet so misunderstood.

Why just today, Grandpa touched her hair as she stood upon clouds forever in view with sweet, delicate songs and the freshest fragrances anywhere to behold. She was not too cold or too warm. Everything so alive, vibrant, pure and sharp in their details was today's adventure. Alycia felt very special and cherished, yet not to her conceit, rather to her comfort of soul. Such was the timeless struggles of dreamers through the ages in this world.

It was always attainable for the dreaming misfit of this 'practical' world, this meander of such lofty perspectives. Generations upon generations have such rare moments, ones so easily dismissed or misunderstood. Not only to be misjudged, but also to suffer from unjustifiable cruelties because they are 'different'. Why these are the ones to lead into the arts and sciences. To encourage observing, measuring and applying the subtler things they experience, that we 'typical, practical' ones not only miss, but attempt to crush in them.

We must allow dreamers their very real 'dreams' … or are they just dreams?

Albert Einstein is considered the ideal of such 'thought experiments'. He was able to imagine, based upon established math and science, a situation that had up to then been mysterious in the physical world. Something that was illusive, just on the edge of the possible to observe and measure at that certain level and capability of science. Oft times, mathematical concepts precede physical discovery. The thought experiment is the assimilation and extrapolation of the data and what it is implying. The trend of this collection of data is then leading to greater understandings, and in the proper individual, to imagine and contemplate even greater things. As H.G. Wells had for the title of one of his short stories, 'The Shape of Things to Come', so let us also try to see some of these 'shapes'.

Such are these shapes within the cylinder of free decision and allowed variability within the greater world line cylinder of the deterministic world line. So much so that the great clockwork of God's Superuniverse proves over the billions or more years of function its own authentication and viability.

We have been extrapolating and reflecting upon the viewpoint of a child. But there are other viewpoints as well; the very old, the autistic, the confused of the present practicalities of our supposedly civilized world and just the plain, misunderstood eccentric. There are so many, varied ways of looking at things, but far too often they are dismissed by common society. Sadly, it has been remarked, if humankind had not spent so much of its selfish efforts defying God, we would have been to the stars 10,000 years ago.

When such things are extremely close or far too distant, they are resolved with less accuracy to actually observe and measure, without finally being dismissed as fantasy by those characters of this world unable to wonder. The rational limits define what is irrational, but it is the counterintuitive, that for some becomes the irrational. That has been the recent revolutions of our state of the extreme mechanics of the also extreme physics herein portrayed.

Chapter 2

The Misunderstood Elderly Man

He was elderly, alone and just one of those way out there thinkers getting by in ripe old age. No one listened much to his ramblings. Most had too many 'important' things to do than to stop and talk to him. In their more concerned, hurried paths of the daily course, they had no time to get involved with him.

Amongst his strewn papers and on his dusty blackboards he would write formulations and considerations that upon closer inspection rendered quite the intrigue of physics and mechanics.

He would often muse, "It is not where am I from, my friend," then lingering with a slight smile, "but when?"

He would sit and rock in his old chair, never in any hurry, comfortable with the rate of duration of the flow of time as it was at its present perceived. All was in order of place and degree of being. All was fine as it was, is and will be. In his past, autism, especially on the lower end of the spectrum was never defined. He had been a dreamer and thinker for much of his years. The curious thing was, even though he had only a certain and defined time, within it he had not been limited in the dimension of time. His ability, as many others who traversed space, would be considered normal. But to traverse time, either forward or backward, was not so normal. To just flow with one's passage forward and rely only on the local curvatures we are residing in, is what all of us do. It takes much more energy to increase the forward duration as time dilation. According to Einstein's relativistic concepts such things are possible. There are likewise some who see an extension to reverse duration as impossible, yet others debate with far more energy that it is possible.

Shuffling around with arthritic aches, limitations and appearing quite eccentric at times, the old man seemed quite content with his life, as if all were quite as expected to be. The very complex details seemed minor to his perceptions of the greater repeating trends.

Alycia stopped by to chat occasionally, and their exchanges were typical of a granddaughter and her grandfather. What seemed so strange were the counterintuitive perspectives of the young child and the old man. They were of different ages and generations, yet neither sensed the incongruity of the wonders around them.

She would talk of her visits in the sky, and he would show her his machines from the cabinets. They would chat and share of things too remote for others to sense, things long ago that had been or were yet to arrive. Theirs were simple, counterintuitive conversations, without mechanism to how they could be, and in a time and place to just reminisce and enjoy again those things so profound.

Time for the afternoon visit to end, and Grandpa had to take his nap. Alycia used to take naps, but was so anxious to see what was over the horizons or what the tomorrows would bring. She had eight years ago just started her life with so much to learn, while Grandpa had already lived over half a century and had learned so much. He had realized how little he had really known and how much more there was yet to be learned, but first just to be marveled at.

They, as other of their kindred, had always needed some time to resettle back to their personal 'here' and 'now' after such rapturous times. The more profound those moments, then the need for more settling down. Their personal relationships would often strain from their adventures. Uncommon difficulties for such individuals, that to meet another similar to them required much time and space. Such lives have been, are and will be very inconvenient for the practical personalities around them. Grandpa would say the non-linear part of the superuniverse was problematic for those who were limited or unable to ponder so counterintuitively.

The child's swing and the old man's machines were of the same construct and mechanics. They were based on the simplest of machines, the pendulum. For in Grandpa's cabinets were the remains of much greater mechanisms, those based on the oscillation of mass. The non-linear back and forth, and the resultant oscillation of the curvature of space are in our normal world so insignificant as to be ignored. But with far more extreme articulations, the most subtle becomes the most profound.

Einstein also said that space and time are connected. And in such extreme states of Hypergeometrical Mechanics, space and time are inseparably connected. Time becomes the more predominant over lesser space, and here so very counterintuitive, so extremely 'irrational'.

Only a child, or an old person with an inner child-like nature, can ever see and hear, observe and measure such extremes of time and space. When the old has the privilege to spend some time with the younger citizens of 'wonders and dreams', then the elder can learn once more how to just go along for the ride. The adolescent learns that they will one day have the solitude to try to figure some of that ride out.

In the very fast, in the faraway distance, in the slowly unhurried, in the closer nearby, there was even more to reality. Events are happening so fantastically outside of our tiny boxes, we have even yet to know.

Grandpa had often reminisced of the time he had taken extraordinary effort to visit that which had ordinarily been considered an impossibility. This time of recollection came after he had some hot tea and lemon and he had napped.

This particular memory was about a parade during a time when brass police and fire department bands would march down the street and play for the crowds lined up on the sidewalks. John

Phillip Sousa was the famous musician of the day in that summer of 1895 AD. The conveyance and the path of his traverse started from 2010 AD. Other attempts had been made to the period between 1889 and 1915 and of such resolution as to remain below the kinetic signal to noise ratio. This to not interfere physically by any kinetic means, and only to receive any weak electromagnetic and mechanical receptions so amplified as to be of interest historically and personally.

The colorful parade, the up-tempo music, the bright sunshine, the slight breeze and the smiling faces were so wonderfully intriguing. The air was filled with a patriotic pride and celebration as little children scampered around with a barking dog here and there. Birds twittered and flew around as the approaching loud marching music increased the tension and excitement.

The fantastic counterintuitive realization of all this activity having been presently perceived in 2010 had already occurred one hundred fifteen years ago. But in the entangled macroscopic tunnel it was again to be replayed in amplification at one hundred fifteen light years away, by means occurring at 370,000 times the speed of light. At a speed so fast, that reverse duration and reversed entropy were the norm, as was the state of not kinetically interfering either in 1895 or 2010. In this state of macroscopic superpositioning, time was preeminent in three dimensions -- entropic, rotation and expansion, while space was compressed to just one dimension. This mantle and substrate of greater measure was then topped by our surface of four dimensions where space is predominate as length, width and depth, and time is but a single, entropic dimension.

Such a profound revelation, yet it is subtle and difficult to normally perceive. The past becomes future, and the future becomes past, the farther in time and space one goes. It is written: "There is nothing new under the Sun", and this is confirmation. Expand the concept to the suns and more distant suns, and 'here' and 'now' means less and less.

At the smallest of resolution, if we could attain it properly, we should see the more transitional and less defined of things. 'Beginnings' and 'endings' are of our own reckoning. We understand time better with them. But go smaller and deeper, then resolution greatly increases and we find further transition and phase. One ending is another beginning, on and on eternally. This where there is less mass, momentum and inertia, until there is none.

No more mass, no more momentum and no more inertia to observe and measure ... gone as vapor. Sufficiently so difficult to observe and measure, that their relevance is as past or future. Now is the only place of kinetic interaction. At this place where the only speed is 370,000 times the speed of light, is when time, relevant to us, is reversed. A backward entropy that is as functional as our forward entropy. If when we are on this side of the Event Horizon all would seem normal, then the side we are not yet on would seem reverse in time and mass-less, momentum-less and inertial-less. This side would appear at 370,000 times the speed of light as well. Both sides of this Grand Event Horizon unreal kinetically to the other, yet each are in need of the other geometrically to survive.

All that is left are the shadows and whispers that existed vapor-like and so non-kinetically, only to barely be detected as something in the noise to be appreciated ... all very subtle, yet so very powerful. Those electrons whirling in their probabilistic cloud around the atom's nucleus, never tire, because they no longer have inertia. They can now be non-local and superpositioning and going 370,000 times the speed of light in their entanglement, wormholes and tunneling.

How amazing it would be to understand ever so slightly the speed of light's limit upon Inertial Geometry; or to cross the Grand Event Horizon so microscopically closer, faster than the speed of light. And then to be able to travel long distances in space and time by either method. Under light speed, time dilates into the future farther and farther all the while covering more and more distance in space. And at 370,000 times the speed of light, it covers vaster distances and goes farther into the past, when one would change their deterministic world line in this superuniverse and not be able to return to their home world line. Such counterintuitive ways are so foreign to our small ways. In under two and one half hours, and one hundred years from now or in the past, the future trip keeping me on my far more tightly curved world line, and the past trip changing my world line altogether. These very deterministic world lines allowing only the minutest free decisions, or rather smaller deterministic lines of sequences of events, than we humans are allowed to tinker with. For on the grander scale, the deterministic trend of God continues. The less the resolution, the more deterministic greater scale dominates. Our free-will is so limited, thus also our understandings.

In about three days one is above the Galactic plane and can view our Milky Way in a greater vista, while at that distance and back in time. For then you are nearly three thousand light years away, and you are the same years ago. If you left on Tuesday, by your Friday you would have been absent from your place and time in history and have lived three thousand years ago. So at this time you could send a light speed message from your Hypersphere above the Galactic plane, and it would be received three thousand years later, yet so weak as to be undetectable above the noise. That signal now within the shadows and whispers to those back on Earth, or better those who were to come in three thousand years.

Worse yet, those who could receive it would not be aware that you sent a signal, except for a prior instruction that you had left behind. Your inertial signal is so non-inertial that it would defy detection. Effectively, you would have no relevant or real kinetic interference. You are for purposes of three thousand years ago, though just a few days ago you were kinetically relevant. Then your past affected their future, and could still, but with such vaporous whispers, that only an exceptional one would have been ready to detect such a signal. At these scales, as a grand exhibition, a signal from your past would be their future. But realistically, a signal from any distance is from the past to the future.

Even three feet of distance is *3nS* in the past. Not as grand an exhibition I may say, significant but not noticeable. So that every day such an effect is not considered strange.

Now let's bring in entanglement and its speed of 370,000 times the speed of light. By the way, it is also the speed of a wormhole. Both develop a chord that not only connects two distant spaces, but also two distant times. These spaces and times are now kinetically connected, to appear 'instantaneous'. This is only an illusion, for at Earth- scale distances of low resolution, this is what 370,000 times the speed of light in reverse time looks like. So with much farther distances in space and time, the counterintuitive is more easily measurable if greater resolution is used.

With entanglement and wormholes we can know that two distantly separated places share the same now and are kinetically very relevant to each other. So our three thousand light years above the Galactic plane signal can be at both places at the same time, because they are in superposition. This is Non-locality and is very counterintuitive. In this scenario the light signal is at the speed of light, and in entanglement already at both places and both times, only within the wormhole.

Before the light beam was gone on the superuniverse's surface at the Event Horizon at the limited 186,282 miles/ sec, the Hypersphere was already within the wormhole or chord of shorter distance to the Event Horizon surface. Now the light beam is traversing through the wormhole of entanglement also at 370,000 times its natural speed.

Now if Earth sends a light signal to the place where our traveler will be at 370,000 times the speed of light on Monday, and he gets there on Tuesday before the signal arrives on Thursday, then the signal went to the future. It is as if they sent a signal about three days into the future and around three thousand light years away. So at one end of this wormhole of entanglement is the past and at the other end is the future, and both are Non-inertially connected geometrically by a Non-local now. Welcome to the past, the present and the future! Two different places are in a sense the same place and two or more times are at the same time in a manner. Non-locality of space and time now have relevance kinetically.

The more pervasive superuniverse is far more fantastic than the universe, simply considered.

Chapter 3

Primal Memories

Oh the chronicles in rocks, in photographs, in audio recordings and in the tracks of events ... such histories are left behind overtly, and some ever so subtly.

So how can one see and hear again something remaining from the past, possibly even feeling or smelling, rather instead 're-experiencing', as per déjà vu? Why, if the theory is quite the practical, would such memories as possibly snippets be passed on to future generations? For instance in behavior, it may be noticed that she holds a tea cup so like a past grandmother who was not well known to the person. Or upon entering an old, unfamiliar building he has the feeling he has been there before. Both examples perhaps are merely subtle shadows of memories from a direct ancestor. It may be that grandfather had also spent time there, long before a person of the present time.

In the New Testament at the end of the Gospel of John, it is mentioned that the Earth is not large enough for all of God's books. Consider that books have pages and then likewise rocks are layered upon the Earth and upon other planets. Why even the fossils are like 3D photos of a certain scene at a particular time in deep history of life upon this Earth. The Sun has layers as do the other stars and these layers though very dynamic, also reveal a progression, a story. So these other 'books' of God are way beyond the capacity of this Earth. We can suppose that everything that has kinetically happened has left some kinetic response in the structures still remaining. A cracked glass window, though no matter how repaired, has some remnant of the original occurrence. On the glass, as for all static materials, the mass, direction and speed of the projectile or better the originator of the trail or track, is indicated quite well.

Being the 'cosmic detective' is quite literally what a scientist is doing by basing evidence on the repeatedly measurable and verifiable. The repetition of the substantiated is increasing proportionately to its certainty, from 0.0 to 1.0 in the mathematics of probability.

By utilizing the experiential aspects of DNA and possibly RNA in the brain, of retaining descendent-intergenerational memories, would be quite the non-kinetically intrusive form of past time travel, short of actually going there. The kinetic signal to noise ratio is so very important.

It is this ratio that allows the present now to be distinguished from the non-kinetic past and the yet to be kinetic future. To measure any shadowy or almost imperceptible signal above noise of past or future requires the best of cryogenics as near absolute zero as possible. According to Maxwell's work, a future signal is as possible as a past signal, but would be very weak. A signal from the past is a retarded signal, and from the future is an advanced signal.

The present now is greater than 99% of the time that the kinetic signal is above the background noise that we observe and measure anything, as an occurrence, or event. But is it really that cut and dry? Perhaps, we are only easily aware of such things as they are presently occurring and so now's fleeting nature is thusly perceived.

In Einstein's Block Time, he said that the past and future are as real as the present. Furthermore, he denied that simultanity was real. Consider that in our everyday lives, things at the speed of light, and even the faster speed of entanglement are far swifter than what we are perceiving and familiar with. This also allows our sensing the just past and present in the world around us, in the nearly imperceptible nano- and micro- seconds.

If we were talking about my saying 'Hello' to you from three feet away, then our present moment seems so 'now'. Your seeing my face and lips move are only delayed by three nanoseconds, and hearing my voice at the speed of sound though much slower, is still in synch that all appears at the same time. Now have me, if it were possible, shouting to you from Mars and the delays would be far more pronounced to decades of minutes. Even if we synchronized to say 'Hello' at the exact same moment, then at the distance of Mars, simultanity would be easily eliminated.

At this point, we now could expand upon the far more intriguing of our conceptualizations herein. What would it be like in the preparation, during and follow through of a mechanism, containing within a human traveler? At such extremes of physical environment, Lorentzian effects begin to take over from the otherwise 'normal' world we have been so used to. At such increasing extremes, what is simultaneous, what is now and present, do become far more difficult to define. Not only in our present avenues of spaceflight, where delimiting day and night straightly is quite the challenge, but in space there can be observed and experienced many different day and night states, upon this Earth and the other planets involved at that particular study in consideration.

The older gentleman previously mentioned has much more involvement with the particulars of Hypergeometric Mechanics. He has quite the tranquil satisfaction of having experienced the extremes of what is permissibly possible. His early submission to basing his conclusions on the repeatedly verifiable, has allowed him to make very startling observations.

Chapter 4

One of Many Vast Journeys

Our paths crossed as if planned. The girl's grandfather and I had been introduced by a mutual, scientist friend. Because of our unusual interests and beliefs, the old gent and I became fast friends. It had taken him those erratic twelve years, the seven tedious, frustrating experimental years, and then suddenly momentous epiphanies in the last three years up to 2010s feeling abandoned by once close relatives, he had entered a more focused time that allowed deeper contemplation.

A wilderness is a difficult place for the creative and reflective. Many times the inspired individual avoids the desert at all cost. But this apparent wasteland could also be a time and place of personal free-will and development.

What loss to bear ... only brother gone, dear mother departed and finally his father passed away ... son and grandchildren so detached. What to do with time ... merely continue working with no plan for retirement, traveling or simply basking in a senior life without any cares for worldly routines of making a living? No, actually the escape into mathematics and physics was some of the most satisfying of endeavors. Being alone, yet reflecting upon God's universe, even locally, but at its extremes, had become wonderment, difficult to share and harder to cease.

Had all of this been in assimilation for decades of the more practical of productive years? His employment in telecommunications, research and development laboratory and the government encoding center were all gratifying. These were all grand places to work, wonderful experiences and to incubate such tangent abstracts to be later actually put into action.

Even the federal facility, the final stage of working life before an official retirement, was a place to think, while avoiding the hired infidels and their unethical ways. This experience clearly demonstrated to him that the only real difference between the empire building superpowers was just geography. Both had their local islands of the mandated corruption of willful ignorance, to the point that less than ten percent were even thinking human beings. The famed 'Roman' pleasures of food and circus had deadened them. In the older gentleman's isolations, rather fruition had started, a blooming that would prove very provocative and fascinating.

The first of the many vast journeys was the most surprising of all. Being prepared only by theory, it was an enormous undertaking. Consider that one can anticipate, but then fall so short of intelligent expectations. Home soon became laboratory and office with stacks of papers, either recalculations following a burst of energy or more inquisitive speculation. Then so weary of mind as to just have to turn away and save for some other day a deeper course into the wilderness of the veiled mysteries.

First it took quite a series of months to articulate the Temporal Diffraction Grating, only such a simple thing to measure the transit of a laser beam through time. It was at the perpendicular to the spatial that time was usually and easily measured. A particularly simple machine was used to measure the spatial as a benchmark for the duration of the time for the beam to traverse its course into the resolution of the nanosecond.

Later the Hyperplane was developed to set the stage of physical environment for the rotation of a pulsed laser beam, kept so simple as to replicate a positive progression in the binary number system as it spun its circular course. Near the axis all was normal as the rotational velocity was ever below the speed of light. As the laser beam rotated at the increasing radius that the rotational velocity was equal to the speed of light, the information's positive progression froze. The detector at that radius had to be rotated in that orbit to differentiate the varying frozen binary renditions applied. Here one was always one, there five was always five. Then at a more increasing radius the sequence of progression was reversed. The detectors measured a regression. The sequence was going backwards as the spinning beam rotated round and round.

This was like a photonic-only Event Horizon artificially orchestrated and quite extraordinary. An information only stream being observed and measured in forward, frozen, and reversed duration.

Being involved with amateur and shortwave radio and always on the study of electromagnetic physics, the question arose, could matter itself be articulated as a signal and not only photons?

To manipulate mass or matter, would be as to oscillate it. This then would be a non-linear motion in a harmonic oscillator. Rather than trying to cover spatial distance linearly, say thousands of miles from beyond the walls of the laboratory, why not in a small space just larger than the essential wavelength required. Intriguingly, this rapid and oscillating coverage of a certain spatial distance would repeatedly suffer the Lorentzian effects of a linear transit, but as a curious pendulum moment, not allowing for dampening. Oscillating faster with such a shortened wavelength that a point hypothesized as foliation or granularity would become predominate in the local space-time.

Cryogenics would be needed for the powerful magnets, based on superconductors in which such magnetic amplifiers would feed from their oscillators. All this just to 'bounce' some diamagnetic material and then eventually a sphere to extreme moments of phase space. The whole contraption was called a Hypersphere, while in reality the sphere itself was the Hypersphere. The complement of the mechanism was supportive only. The disk under the levitated sphere was the driver, as in an antenna, while the sphere was the 'signal'.

Only small diamagnetic particles had begun as such 'signals' and evolved into spheres of varying sizes, until a sphere of the proper dimensions that could comfortably house a human being had come to be.

Taking over more than a dozen years, this was finally achieved by this mechanism with only one moving part. The Hypersphere was something quite unique in its more polished and finished state. So simple was its design, yet so profound its function.

During 2010 was when so much had been accomplished, written down and archived; yet silent was the return upon its theoretical and experimental papers. By this time much was in process to be sent to Israel, Great Britain and the United States. It was certainly interesting, but perhaps not well understood. While some were quietly letting it lie dormant, the older gentleman had already gone quite far with its potentials.

By 2013 even the Prince of Wales had a hard copy about this Hypersphere and had sent a private letter to the older gentleman. All was subdued, while very few were aware of what had been going on.

Only a few understood that to traverse great spatial distance in far shorter time would mean that reverse time had to be anticipated. Also the same few knew that to traverse one hundred light years in under two and one half hours was to travel there when it was one hundred years in the past. The first by entanglement and traversing at 370,000 times the speed of light. It could also be traversed to one hundred years from now into the future by getting close to the speed of light. What was amazing was that at the loss of inertia no longer were acceleration or deceleration required, for one was already at this 370,000 times light's constant in velocity.

Greater distances were feasible, but increasingly inconvenient. At this time 1895 AD was one hundred fifteen years back in time and just over two and one half hours away. The same but forward in time for 2125 AD. The danger was in losing the entanglement, thus changing world lines in cosmic history in this superuniverse we dwell in. One would not come back to 2010 AD, but another that was the forward progression of an alternate world line of history. At this home world line, one had just disappeared to never return.

The implications of the application of theoretical geometry to practical mechanism are provocative.

Utilizing liquid nitrogen to cool a superconducting disk was already developed, and the test was with small particles of pyrolytic graphite, and later on bismuth. Both of these types of particles are diamagnetic which allows a counter magnetic field to form in polarity to one induced. Thusly, they would bounce in response to the pulsed magnetic field part of the oscillating cycle and then fall according to their mass in the Earth's Inertial Geometric during the next part of the oscillating cycle.

Back and forth as a pendulum, on and on this Harmonic Generator, the Hypersphere, would proceed. Faster and shorter would the frequency and its appropriate wavelength equaling the speed of light, go. Articulation was ever so slightly 'forcing' the wavelength to stretch to allow for the consequential Lorentzian effects to occur.

Faster and just under the speed of light, would give a positive or forward time dilation; at a certain point 'slippage' through the granularity or foliation of space-time would allow negative or reverse time dilation. Interestingly, in the reverse total loss of inertia, Non-Inertial Geometry occurred and thus suddenly without the need for acceleration or deceleration. Just as suddenly, the particles or 'matter signal' were now at 370,000 times the speed of light, so as to race across great distances of space of forward or reverse time in an entangled state of superpositioning.

This occurs astonishingly at somewhere around 90% to 99% of light speed; but only with the particulars of manipulation of pulses during oscillations between a region where microwave, terahertz and infrared to red occurs. This is very significant. Depending on such a local frame, dragging phase space non-linearly in this mechanism of pendulemic mass in such a harmonic oscillator, allows for such great things to happen. In the macroscopic scale with instrumented spheroids, quantum and relativistic non-locality and tunneling had been achieved. These particulars would have been already sent to Israel.

To travel one hundred fifteen years into the future, one would suffer great increase of mass, excessive foreshortening and time dilation for nearly two and one half hours, while the normal world would very slowly pass into the future, still in the Inertial Geometry of so called gravitation with very compressed inertial and kinetic effects. To go much faster though, one could enter a Non-Inertial Geometric at 370,000 times the speed of light, no longer with observable or measurable mass and literally not of this world; and at some point in space one hundred fifteen light years away, while one hundred fifteen years past or future in time, without kinetic effects, because of no inertia and not really 'here' anymore.

Thusly when one's kinetic signal to background kinetic signal to noise ratio is low enough, the present and now do not mean anything. One is no longer here, but either was or will be. In the quantum world it happens all the time with so called 'virtual particles', but now in our macroscopic world.

We too can be Non-local as per Bell's Non-Locality Theorem, and understand how real in this superpostioning state, past and future are as real as, or more real than the present or now, and this kinetically. For it is only in the kinetic that things can interfere with each other in our decoherent state so we can interpret here and now and the present.

How terrifying the first time it happened must have been. Then with continued effort and far more preparation, would the fearful become wonderfully fascinating? Space and time were interactive, and astoundingly so. The older gentleman's imagination had considered one afternoon a parade in 1895 near his maternal great-great grandfather's little store. Not to interfere, just to observe a time, supposedly gone, yet if remaining entangled not to break his present world line. This would sever his return to this world line of history. He would just disappear going on in another non-linear sequence of events of an alternate world line. Our original world line would continue on, but our adventurer would just have vanished at a certain date and time and would be unto us gone forever.

The summer of 1895 would be a temporal target to experimentally try to entangle and tunnel. The location would be on Market Street in Amsterdam, Montgomery County, New York in the United States. The famous Fourth of July would be quite the scene. The more specific the time and place, the far more accurate the articulations required. All of this from 2010 AD in the exact location in Amsterdam, NY in the United States to a target in time one hundred fifteen years earlier and one hundred fifteen light years away. The accuracy and precision was very demanding.

This must consider the superuniverse as a rotating sphere, even a rotating super black hole. The Event Horizon presently the two way door and the surface as the velocity of the speed of light.

When entering the extremes of relativistic, forward time dilation with the increase in mass and the foreshortening of length as per the axis of oscillation, when the 'slip' occurs at a very precise point, one is suddenly with inertia lost and at 370,000 times the speed of light, a tachyon. If just below this point ever so slightly, one is still with inertia and forward in duration, not a tachyon. If just at or over this point one is a tachyon, in reverse duration. At just under extreme, one is coursing to one hundred fifteen light years away in the forward direction, still on this side of the superuniverse's Event Horizon, in non-linear relativistic spaceflight.

At and just over the extreme point, one is coursing one hundred fifteen light years away in nearly two and one half hours, but in the reverse direction. Likewise in one hundred fifteen years forward into the future it also takes nearly two and one half hours

In the reverse state, one chords through the superuniverse's Event Horizon so as to tunnel to the past on the other side of the Event Horizon. This is inverted or negative, relativistic spaceflight. Time is reverse to our side of the Event Horizon.

All this just described is very Non-linear, while up to now most travel on Earth and in space has been linear, such as a train or a rocket. Even relativistic particles being accelerated at Cern, Switzerland followed the same response. In the Non-linear it is pendulum-like, a harmonic phase space articulation within the oscillation of diamagnetic mass.

If careless, this manipulation is catastrophic. When done very slowly it is entangling and tunneling with stability. When allowed to run increasingly wild, then it is like matter and antimatter, with the 100% trade of matter for energy, at the Event Horizon. The sudden release of an immense amount of energy traded for the matter now on the other side of the superuniverse's Event Horizon, approaches slowly or tragically anywhere in our rotating superuniverse by 'slipping through' or crashing through its Event Horizon. The atomic weapons of our day, trade around 3% in fission and 5% to 7% in fusion, with very powerful results.

That 'surface' is the velocity of the speed of light, around 186,282 miles/s. A harmonic Non-linear oscillator is near the speed of light, and just for a precious moment. Its absoluteness is now quantum-like and not so discernible. We are at Coherence and under those strange quantum laws at and in this strange Non-inertial place. While relativistic velocities below the speed of light only approach it, and still retain inertia, yet showing extremes of Lorentzian mass, length and forward time duration effects. Just near entanglement and tunneling and onward to the future in a compressed relativistic state of mass increase, very shortened length in the direction of motion, and very slowed time still forward occurs.

When again at suddenly 370,000 times the speed of light is the loss of inertia. This is also the speed of entanglement and tunneling, and in a reverse entropy of time duration, then gone from this world, and now past.

This is the older gentleman's Hypergeometric Mechanics, operated so benignly, so artistically as to be non-intrusive as possible; yet as before, entangled so as to return to this world line, his home space-time. Those around him share this history while he witnesses it as the replaying of times so past and supposedly 'gone', almost like a vacation or holiday, and quite refreshing.

The smells, the sounds, the warmth of that past Sun and its light upon the surroundings, the dust of the street's activities and all the recreation of the events of that day and place were mesmerizing. So little of the available electromagnetic and acoustical-mechanical energy he would observe and measure, as to have only the most whispering and shadowing affectations he would need to utilize. He was of the most flimsy of interference kinetically. In any practical sense, he was almost, just almost there.

He considered the camera, as well as any recording device, a quantum machine. For the allowance of a little extra energy, as a light to see an old photo, the past moment was portrayed. So even he, with his eyes, ears and brain was a quantum machine in an entangled quantum state, but now on the macroscopic level.

The question of determinism on the grander scales and free-will allowances on the minor scales of space-time, for him had already made sense. Perhaps as the greater flow of the course of events had within it far smaller, flexible courses that swayed side to side, that also flowed in the same direction. The smaller courses would represent our limited free-will within the far larger confluence of the determined course, all in the same cumulative direction. The much smaller has angles that could represent the percentage of free-will in sympathetic flow nearly to the determined grander current.

The grander current, thusly could be within the world line of these greater non-linear sequences of events inherent. As a decision in nearly parallel angle, say near ninety degrees, then such would be far more possible. A decision far to perpendicular, near the zero degree angle would be extremely rare, I daresay almost impossible.

Such would be the gentle old man's discourse if one would allow him to rattle on until some would succumb to a nap while perhaps others might be inspired with understanding as to be enraptured to the profoundly provocative. "My God is such the Maker of worlds," St. Augustine suggested, to which he would always add, 'and the destroyer of such!'

The old man was getting tired after our visits and chats of such wonderment. He had begun to settle in his soft chair and doze off to sleep. I had to get some practical things done that dealt with the more familiar here and now. But as I was leaving I noticed a drawing laying on his table of the constellation, Cassiopeia. It had an extra star to its west, as one would look upon it in the northern sky of night. I looked closer and it was marked, 'Our Sun'. Upon further inspection of this interesting astronomical arrangement of the stars, it was also noted, 'the view from Alpha Centauri'. It was so fascinating to contemplate so greatly for humankind, but so cosmically. Our stellar neighbor possibly in co-rotation around our Milky Way and journeying with us through past human history and earlier.

Chapter 5

To Alpha Centauri
At 370, 000 Times the Speed of Light

In the night sky nearer to the lower latitudes is the triple star system, Alpha Centauri. It is 4.32 light years away and made up of three stars -- Alpha Centauri A, a yellow main sequence star, a bit older than our sun; Alpha Centauri B, an orange star about one billion years older than the Sun; and Proxima Centauri, an M dwarf, that some feel is only passing by and not connected to the other two, or is in an extended orbit around both of them. If connected, it too would be about one billion years older than our sun. All of these are possibly from the same open cluster as our sun and conceivably siblings, as their chemical signatures are similar. Being the elder siblings may have allowed them to progress ahead of our sun's family of planets.

It takes light and other electromagnetic radiation 4.32 years to go either way, thus a round trip is about 8.6 years. For a 10% speed of light spacecraft that would mean forty-three years one way or eighty-six years round trip, and that is not counting acceleration and deceleration factors.

If, as the old gentleman relates from time to time, entanglement and tunneling, occur at 370,000 times the speed of light, but in reverse time, then a trip to and from Alpha Centauri would be just minutes away. Minutes, yes, but with interesting consequences ... the spatial traverse is in reverse time!

I remember him mentioning about one and one half minutes for a light year and so about six minutes to Alpha Centauri. And yet, I merely thought he was hypothesizing. Or was he? From another one of his papers he had mentioned to me that it was about ten minutes to Tau Ceti, so that agrees with just around a minute or so for one light year, since Tau Ceti is about ten light years away.

He was such an entertaining and imaginative old man. He was probably just sharing some of his fascinating thoughts with such cosmic distances and the universe with me. He had such a smile, as a Cheshire grin.

He spoke with authority as if he had already experienced such dramatic conceptualizations. I had to wonder and often would just bring myself to ask a question so begging to be answered. Hesitatingly I was finally able to ask, "Sir, what if there is such a machine?" To which he simply sat for a few quiet moments, then that smile reappeared.

He replied, "What if there is a machine? What if?" His eyes seemed to so loudly fill in the implications provoked.

"There is a little girl who could answer that for you. I dare say she understands though as a child here and now, for she has eyes that see and ears that hear as only a child can," he paused and then continued. "I still wonder, then think, but I ponder too. It is quite the pleasure. Her name is Alycia.

As I write this journal in my own old age, I remember an old woman that had visited. Her name was also Alycia. She had come on one of the local country roads near town where she had the driver of the vehicle pull over while she took a walk into the nearby field. There used to be a house there. The field was very significant it seemed to her. I remembered she had told her own granddaughter about the Golden Cites of the sky, and where the swing was.

Alycia had a diary journal as well and had given it to her granddaughter. In another entry she had mentioned how she not only was able to visit and observe her great-great-great grandmother, but also her great-great-great granddaughter. It is written that we have a 'great cloud of witnesses' ... I sense not only in space and time of the here and now kinetically, but subtly of far away and in the past and future, non-kinetically enough so that there is no relevant inertial interference.

When Alycia met her grandmother, and also when she met her granddaughter, she was familiar to each of them as such a person they felt they had 'known', and as amazingly implied such in her diary.

Of course it is common agreement that those events situated in quantum and relativistic extremes of physics are counterintuitive. It is the extremes of such geometries, such as Hypergeometries, that allow things so subtle and vaporous, so abstract and esoteric, to be so predominate. As I related to such scientific speculations and numeration, my curiosity was intrigued by a journal on the old gent's table. You see, while he dozed I would flip through its pages. The drawings were quite fascinating but the writings were not so easy to process. Finally on one page I had passed over many times was the faded caption, 'The Hypersphere'.

Had he actually built such a conveyance? If he had, was it of any use? Surely not to consider practical for most purposes, I am sure. But of the extremes for which it was intended, it seemed quite adequate.

In a jumble of notes, he had dissertations upon a world with two suns. One was much like ours and the other more of an orange, one a G class and the other a K class on the main sequence of stellar progression. Many of his notes were faded and for these I would have to have some stronger light or even shine a light through the paper to better make out what descriptions he had written. It was somewhat chaotic as to page sequence as well, for the numbers were so pale as to be missing.

The planet though described was a bit more massive than Earth with more ocean than land. Large deserts were on those continents, while the islands and archipelagos abounded with dense plant life, megaflora. Hesitated in my perusing, I considered that such large plants would symbiotically require large animals, megafauna. The gravity, because of the larger and slightly denser mass of the planet, seemed to qualify broader and more stunted designs in these living things he documented.

On and on I would read until a snort of awakening would signal me to be prepared to return any of his literary work back upon the table. One time he awoke and just looked at me and smiled, apparently satisfied someone bothered to read his work. Then off to drowsing again, and I slightly flushed with embarrassment would go back again to where I had left off.

One area which caused me to shake my head and take a second look was when he referred to ruins. He denoted very large, plain stone upon stone structures in the near desert areas. For the more gravitational, the old man conjectured a 'tighter curvature of the Inertial Geometric' effect would require quite the muscular strength and also any engineering brutishness as well, to construct such what were now ruins.

This all is so astounding in itself, but if this was not science fiction, then he would have had to travel to and from Alpha Centauri's planet, about six minutes each way. Not bothering to consider how long he may have stayed, for that is a whole other story.

About other planets in the Alpha Centauri system I so far had not found anything, but I imagined there were others. Modern astronomy records of Alpha Centauri A and B, two co-rotating stars about eight astronomical units apart; the G star is yellow and the orange K star.

A third star, Proxima Centauri also seemed to share with the system at a much farther distance and was a red dwarf, an M star. Again, this system according to some astronomers and astrophysicists had been possibly co-orbiting with our sun for hundreds of millions of years.

If there is anything to panspermia between planets, it is also of a consequence between stars over much longer periods of time. As I have mentioned, the longer the time and the greater the space certainty arises over more local rarity in time and space.

Even keeping one constant, either time or space, there is a rise of certainty of a lesser factor, while the other increases. Thus, with both increased the implications become astounding, which made me quietly grin in delight and fascination.

As the old man stirred, I did not want to wear out my welcome, for I found our chats so interesting. He was such a quiet yet intriguing character.

"I need to get going, my friend, for I have some daily errands to do. May I visit you again soon?" I asked as waiting for permission, already knowing he would be just as anxious as I.

"My, yes, indeed, yes indeed!" he replied as he suddenly lit up in countenance, the child within so prominent now.

"Good day then," as I let the door so quietly close behind me in response.

Chapter 6

The Past to the Future
The Future to the Past

As a young man, it was the loss of loved ones that prompted his exercising alone, putting into experiment his theories. It was Albert Einstein's Theories of General and Special Relativity, Kurt Gödel's Theory of a Rotating Universe, and previously H. G. Wells', *The Time Machine*, that had always fascinated him.

The tragedies of the passing away of those close to him, all in such a short time were effective in focusing him into the predominance of time. He could spend days upon days, except for 'making a living', studying and experimenting ... no one to fall to the side in frustrations to distract him. Quantification upon quantification, soon concepts began to make some pattern of function. Increasing of resolution offered more subtle granulations of the fabric we call space-time.

In time, it was space compressed to one dimension, the time expanded into three; one that which we are familiar, increasing entropy, the next two quite surprisingly, rotation and expansion. The temporal dimension had a similar, reverse function just on the other side of the Event Horizon. On this other side, was reverse entropy, reverse rotation and contraction. When one would observe from the other side it seemed as forward and normal as our side does. From either side, the spherical, convex curve of the local Event Horizon equals the superuniverse's Grand Event Horizon and is just its local anomaly. So under certain extreme conditions, I had aforementioned, extreme actions lead to extreme reactions, as Sir Isaac Newton so well stated.

What if there were such a machine? Whether there had been or is, or will be, the capability of such a machine is equivalent in Einstein's Block Time. Time is as traversable as space.

If conditions are right in God's local part of the vast superuniverse, even a little girl on a swing or a young man oscillating in a Hypersphere, one by innocence and the other by design, can journey as so few experience.

The past reveals and can be a gateway to the future, as the future reveals and can be a gateway to the past. It is the essence of strategic planning for things to be, as well as a detector for things that were in the past. If the inertial can transcend to be Non-inertial, at least for a time and place, then the now and here become irrelevant for a while. Irrelevant in that Non-kinetically interference occurs between material objects ... similar to Bose-Einstein-Condensates occupying the same place, coherence in Quantum Mechanics versus decoherence, and tunneling versus being unable to pass through 'impenetrable' barriers.

Consider those Golden Cites in the Sky and the parade in 1895, so counterintuitive when all is not in the extreme and quite the norm in the extreme. Go fast enough and one never arrives in the here and now at a wall, but is suddenly on the other side and traveling in the reverse time along the same direction; or in the opposite direction in the same forward time. Surprisingly such feats are presented for our academic consumption. Perhaps, we are just limited in our concepts of what the limits of calculus are saying, along with the imaginary numbers, so often considered just imaginary, but are not.

'To be or not to be ...' and 'though this be madness ...' even Shakespeare spoke in such wonder. From a linear pursuit at high speeds, replaced by Non-linear oscillations at high frequencies ... 'to be, and then not'.

If such local conditions are conducive, a little girl can suddenly without much mechanism but for her pendulum swing then disappear. For her to find herself soaring into the sky, but to others below seemingly oblivious, yet she sees a new world. Enraptured she is there but without apprehensions, free of harsh restraints upon her wonderings.

On one of these flights of dance, while waiting for her mother to have dinner ready and with homework yet to do, Alycia had for what seemed only minutes been unavailable. Her mother looked again at the empty swing and frustratingly called for her dreaming daughter.

"Kaylie, I don't know where your sister disappears to, but she and I are going to have a serious talk about life and responsibility," said their mother.

Kaylie, Alycia's younger sister could not put words to what she could grasp of her sister and her strange behavior. So she just smiled and picked at Alycia's food, while their mom exasperated herself with the easier, horizontal world of practicalities.

Gone for only a few minutes, Alycia's absence was actually a short time while to her it seemed like hours. She had done this before and was able to soar into the sky of the setting sun, the clouds so golden bright and inviting. In the cities she could even hear music and saw the many inhabitants while seeming to fly between the cloudy pillars so gold and shining. It was more daydreaming than sleepy night dreaming, so wonderful to go somewhere where she could visit with those of long ago.

There were many cities there, and though in a strange place, it gave her no sense of fear or foreboding. It was so calm and alive there, more alive and real at times than back on the swing in the practical world where such flights of fancy were scorned.

Kaylie had listened often to her sister speak of these Golden Cities in the sky and other things that Alycia would chat about. Kaylie wondered too. One day while Alycia was with her mom at a friend's house and her daddy was sleeping on the back deck, Kaylie went on the swing. She liked the different feeling and was seeing more clearly, when her dad woke up and said, "Kaylie, be careful on that swing. You're gonna fall off and get hurt and then you'll be in trouble."

She slowed down, got off and went to play in the dirt. After dinner, Kaylie was in her room looking at her coloring book. Alycia was staring into her mirror. She seemed to like to look deeply into her mirror. She had read about Alice in Wonderland, who went through a mirror. It sure was an intriguing idea. But Alycia was puzzled why she could not. Kaylie watched Alycia give up and run downstairs to watch cartoons on the television.

Kaylie waited just a moment and went over to the mirror. But she turned out the lights, closed the door and just let enough light and sound in as to feel alone to think and wonder. She noticed that the hall light just outside the door was a single bulb. In the dimly lit bedroom, this sparse light shown upon the opposing wall making a series of light and dark lines, similar to an interference pattern. Kaylie moved the door slightly opened and then slightly closed, back and forth watching as the light patterns changed. At just the right opening of the door, the interference pattern, those bright and dark lines really stood out. It made her wonder how they were displayed. The brightness increased to the middle of the brighter lines and the darkness increased to the middle of the darker lines.

She wondered if they would get brighter depending on the amount of the incoming light. The darkness never seemed to go completely dark either, but did get darker if she used the dimmer that daddy had put in their switch. When she closed the door it was quieter, but not totally silent, and when she opened the door it was noisier depending on the downstairs' louder sounds.

Her grandfather, Pipa, had once mentioned something of the 'silence' in between the silences, but yet there was never any complete silence. He also had said that the lights were always more bright toward the center of any bright line that had this increasing characteristic, depending upon the incoming light. But the same was for the supposed absence of light as for the absence of sound; there was still some light and some sound, until one would need special high resolution instrumentation to measure it. Grandpa Pipa used big words sometimes, but tried to explain them too.

Grandpa had even used liquid nitrogen to cool his detectors on an early experiment of his. Kaylie remembered some of this, but not enough to really understand it. This was the Temporal Diffraction Grating, which he called the 'Where Is Now Experiment'. He too had noticed similar things as a child, and more so as an adult. He wondered for he was a natural scientist. Thinking was important, but never to give up wondering as a child; this he excelled at to such an extent that he seemed to not care so much of the practical side of things. Sometimes in the mundane he appeared absent-minded but not with the latest technologies.

What held his interest was not of the commonplace as most people, but of those things that indicated greater thought and imagination, mathematics driven by observation and measurement. The implications of what were almost less perceptible than shadows and whispers for the more 'sophisticated' of humankind; of those things possibly far more important to the structure of the universe. That structure possibly more foundational than what was the more easily known.

Alice in Wonderland had a mirror and Kaylie had a mirror too, but hers was an 'Inertial Mirror'. Pipa had mentioned it once and at that time it sounded too hard to understand. It was the same for 'Inertial Buoyancy' as he would re-define what in celestial mechanics is known as an orbit. In the explanation of terms a geometric to Hypergeometric perspective can be included, along with a predominance of time over space, and finally allow the so called 'imaginary numbers' to be more real. Then with such concepts far more seriously considered, it allowed us to understand deeply and offered new regions to be explored. The so called 'counterintuitive' was now not so bewildering and in such a better clarity as possible. With such fresh vistas, new worlds were offered to glimpse that were considered only fanciful before.

In Grandpa's laboratory, a couple of Newton's Cradles were on a table. These often used by others in the world around him as novelties in offices and parlors to occupy one's pensive moments, while supposedly more important things were reflected upon. They were of very particular importance in the journey from Inertial Geometry to Non-inertial Geometry. In his younger days, Pipa was also interested in the studies of matter and motion.

James Clerk Maxwell in the nineteenth century reflected upon matter and motion. At that time in his younger days the pleasant old man found these five spheres of a group pendulum were exhibiting fascinating concepts. Whether a single pendulum or as a group, all were Non-linear oscillators. Phase space concepts of the seventeenth century Hamilton and all that geometry of inertia now in performance fulfilled. Here were to the old man such inklings of things with only the pendulemic movement interacting between their masses in motion. All this possible only in what many call a 'gravitational field', which the old man of deeper resolutions redefined as the 'curvature of the Inertial Geometric'. The sharper this curvature, then the stronger the gravitational field per say.

"It is all so Non-linear," he would say to Kaylie to intrigue her bent to understanding, while she only read his fascination with a smile.

As a single pendulum or in a group oscillates, it gains energy, then loses it and then gains some of it again, however each time with a very slight loss or dampening. As a passive oscillator, this would go on as it slowly lost momentum and distance of arc, and gradually slower in speed. Ultimately, it would stop at the bottom of its energy well and be still.

Now if it were powered somehow or made an active oscillator, then it would no longer dampen, but continue to properly repeat the arc articulated, over and over, as long as some push was reapplied. Thus it is an Inertial Mirror oscillating back and forth. It is as if mass is now a signal. The mass is oscillating, as in a laser to and fro. As in a laser, the photon is oscillated back and forth, but amplified to a point that it proceeds out of the laser as a beam. Thusly a mass can be so articulated in an active Inerital Oscillator, now amplified as well, but with one of two results. If articulated by one side of the push and then a push the opposite way, continuing then it would break free as a projectile. Quite an effective 'Inertial Cannon or Gun'. And if restrained appropriately and oscillated faster and faster, our Inertial Mirror begins to produce increasing Lorentzian effects. These effects are mass increase, length foreshortening and time dilation which are slowing the rate of duration. Restrained enough, the oscillations hit a saturation point where at that moment almost imperceptible to measure, it is gone. It has passed through the Inertial Mirror at 370,000 times the speed of light. It is no longer inertial, no longer kinetically interfering with the still and quiet 'normal' world around it. It no longer requires acceleration or deceleration, for it no longer to our normal world has inertia. It no longer is restrained by the local curvature of the Inertial Geometric. At 370,000 times the speed of light, it has traversed one light year in approximately 1.5 minutes, and the distance to Alpha Centauri in around six minutes!

It has not traveled in our geometric space and time though. It has 'tunneled' through a chord, slightly below the 'surface' of our superuniverse's Event Horizon. This 'tunnel' is in reverse time and to us is occurring at 370,000 times the speed of light, yet to us disappears. This is the 'mirror' that Kaylie would pass through, so much more involved than Alice's.

Any distance traveled was always now in reverse time to the original state of normal after passing through any Event Horizon of our universe, which is really only local apertures of the great superuniverse's Event Horizon.

Kaylie would often look at Pipa's drawings of a great sphere of rotation of the superuniverse, of many universes which appeared as flat islands of circles floating on this great sphere. Kaylie heard Grandpa say that as the superuniverse is drawn as a fifth dimension of Inertial Geometry, the four dimensional universes indicated were now reduced to flat circles. All of this surface being now an important velocity of light itself, around 186,282 miles per second. It is the photon now massless, having the only thing left to be affected by Inertial Geometry's curvature, momentum. It is thus the edge of our inertial world and the edge of the Non-inertial world, the world of tunneling, entanglement and reverse time, even for properly articulated macroscopic objects. These objects were no longer tiny subatomic particles but perhaps superluminal observations of cosmic scale anomalies. Then could everyday objects such as baseballs, pendulum movements in extreme active oscillations and little girls also articulate this way?

It would be some time on a summer's day when Kaylie was visiting Pipa and wandered into the strange sphere in his laboratory. Below and surrounding the lower part of the sphere was a disc that was the magnetic driver. All one had to do was climb up and go inside.

"Kaylie, where are you?" queried Grandpa as he began searching for Kaylie. "Pipa has some lunch for you."

"I'm right here, Pipa. See I am in here," Kaylie said with a wide grin.

"Well, maybe you would like to take a ride sometime with me in the Hypersphere?"

"Is this a Hy-Hy per ...?"

"Yes, it is my dear and it is quite a different kind of machine," Grandpa replied picking Kaylie up and walking from the machine. They enjoyed their lunch together. The old man was grinning with satisfaction at his granddaughter's interest in his ideas.

At the table on a sheet of paper, he drew a circle. "A long time ago before I was born, a man named Kurt Gödel considered the universe to be rotating. And to just travel along this rotation fast enough, one could go to not only the future, but also the past," he instructed. Then he folded the paper up and put it in Kaylie's backpack to take home, realizing she did not understand what he was talking about, but that she would have something to keep that he had given her that day.

It was early one morning when Kaylie was lying on her bed looking up at the ceiling. She was just wondering and trying to remember all that Pipa had said. Some things seemed to be very easy to 'see' while others were so complicated that she fell back asleep from trying to figure out everything he had told her.

Was she dreaming or did it happen? Sometimes dreams are so very real that after waking it can take some moments to reorient to the everyday around us. This time she remembered going to the mirror's inner room and passing through it. She remembered a place that looked very old with horses and wagons and voices and sounds. The smells of cooking too, just as Grandpa had experienced. It seemed that she was walking outside and saw another little girl standing there getting her picture taken, but the camera was very old and everyone was wearing very old looking, strange clothes.

Kaylie's mom came by her room, looked around and discovered the little girl's bed empty. "Kaylie? Kaylie! Where are you?" Her mom searched now as if the little girl was hiding or perhaps just out of view. Then she became concerned and started to look around as if her disappearance might be more serious.

"Kaylie! Mommy is looking for you and you need to answer me," called her mother more urgently.

After going back downstairs searching the house for a panicky half hour or so, she rushed back upstairs.

"Kaylie! Where have you been?" sternly asked her mom. "I have been sick looking for you and I do not appreciate your not answering me. Where have you been?"

"Mommy, I was right here watching another little girl get her picture taken", Kaylie innocently explained as she wiped her eyes and then yawned.

"Oh you were just dreaming, but I could not find you in your bed. Now where did you sneak off to?"

"No, I was here, but the little girl and the people there were very old looking with such strange clothes and everything was so old ..." she trailed off as she fell back onto her pillow.

Slowly, Kaylie's mom rose from the side of the bed and turned around. For a moment her eye caught an old photograph of her grandmother. Earlier that day, it had gotten Kaylie's attention too.

The framed picture had always been there, but today Kaylie had felt that she was there to watch her grandmother as a little girl get that photograph taken. She remembered a small dog in the arms of another woman, but it was not in the photograph. About a week later Kaylie brought up this 'dream' to her grandmother's friend,

Emma, who lived down the street. Emma was surprised about the dream and that Kaylie had seen a dog. The dog was 'Puffy' a childhood pet of Kaylie's grandmother. Emma thought for a moment and offered to Kaylie a question.

"Kaylie, what color was the dog?"

"Um … I think it was small and white and had a red ribbon around its neck."

Emma was quiet for a moment and then she whispered aloud, "That was about 1910." Then the older woman just closed her eyes, looking afar off with a quizzical grin upon her lips.

Chapter 7

Where When is and
Wonderland

As this time forward duration had passed (that most of humanity lives through at the rate of approximately one second per second at the average local curvature of our earthly normal existence), everyone has progressed in the days of life. The girls have matured and become adults. Their grandfather, Pipa, despite some temporal oddities has also aged and become geriatric. History continued to unfold and civilization, as 'civilized' as it fools itself to be, continued to evolve. Whether all progress is really progressive has and still is up for interesting debate.

As for the years past and the years to come, there is the continued Non-linear sequencing of events. The rate of time is linear, as long as the local curvature is kept the same. But events seem to always form clusters. One only has to watch traffic flow to see that a car goes by, then a few, then no cars, then five and other apparently 'random' amounts. Then with the passage of twenty-four hours, clusters form that would under analysis indicate morning and evening rush hours, and breakfast, lunch and dinner activities. Over time even holidays would surface with a multiple year series of observations and measurements. Thus not being able to precisely consider when a very specific event would occur, the ability to sense the trends of clusters would be repeatable. This all could be done as well at a coffee shop or better a restaurant. Daily, if one sits and just data logs the apparent 'random' number of people coming and going through the front door, over a large enough period of time, one could deduce the same repeating cycles of human behavior, of eating and working activities, and holidays. The seasons, and more over time with increasing resolution, even far more subtle cycles of repetition appear.

The question of how much 'free-will' we have seems to beg to be asked in a more determinant universe, that the rationalists hypocritically avoid. If the math is of sufficient duration of data over time and human history (as well as geological, paleontological and cosmic histories) with the minimal cycles of free-will decisions here and there, then along our world line we have only an ever so slight influence.

It can be likened to a pipe with one direction for the flow of history, and a small variance allowed beyond the resolution of measurement, as a thin straw therein, where we have such arrogant illusions of our supposedly 'great' decisions. It appears we have some, but we are less aware of the general flow of things and just assume our perspective of 'importance and control' is so grand.

Consider all these world lines flowing in one direction on a Riemann Sphere from one point to an anti-point. Then rotate all this upon an axis through these two points. One point can refer to 'zero' and the other to 'infinity', both like 'random' are only admitting our inability of resolution to measure. Each line, another series of 'infinities', represents a single world line flow of history, a series of 'whens' upon a Riemann Sphere of our superuniverse. Perhaps each is a world line of its respective universe, and within, are many more 'ad infinitum' world lines, but all flowing in the same direction.

Again the surface is the speed of light and below this surface is this Great Event Horizon, the 'other side' of the superuniverse, in reverse time. Our perspective of all this is as a sphere with a convex surface while to the other side, it also appear as a sphere with a convex surface. And to those there, it would look normal in forward time and we would appear as reversed time.

Consider now this Riemann Sphere, or just sphere, of five dimensions, the fifth being Inertial Geometry, or gravity. The sphere to each side is expanding and each side only sees their time in forward duration. Remember, we to the other side, beyond inertial observation and measurement, are seemingly 'non-existent'. But if one could see the other side, then each side would consider the other side in reverse time. So to better explain this, each side of the Event Horizon's 'surface' is in opposing time flow.

Therefore time has three dimensions ... duration, rotation and expansion. The Riemann Sphere is expanding, but the opposite appears to be contracting, and the Sphere is rotating in opposing directions to the other. Now we are spatializing the temporal and have reduced space to one dimension, as the temporal is normally viewed.

This Riemann Sphere then is a model of the superuniverse, with the side we exist on as in forward duration or entropy. This expansion thus explains the supposed 'Red Shift' as our universe slightly bends over the superuniverse's edge of its great curve. Rotation, as all things seem to do in the cosmos is of prior cause celestial mechanics.

Now to journey to these extremes, is where, or better when, geometry becomes mechanism. 'When' and 'where' as concepts are exchanged, yet all continues to function. "The application of theoretical geometry to practical mechanism is provocative!" has often been said by the old gentleman. His lectures and even homey conversations would also diverge from the here and now, to the past and future. He appropriated Albert Einstein's consideration in his Block Time that the past is reality, just as real as the future is reality.

So what or where or when is 'Now', what we call 'the Present' or the 'here and now'? With causal consideration that all is Inertial, it leaves one in quandary or speculation without enough substance to pursue. But if now, the present or the here and now can be reduced to only that which objects can 'kinetically interfere' with another object in the same place and time, then we may have a better definition. This would also define the past and the future as areas which cannot interfere with each other kinetically or with the now or present. This does not deny their mutual existence, only that they cannot kinetically interfere, because that requires inertia. Without inertia and its kinetic interference, our ability to observe and measure is just an idea.

It is like the fine spokes of a very rapidly spinning wheel, moving so fast that if we rapidly place a very slim stick or our finger, into the spinning spokes, there is no relevance of the spokes of the wheel to the stick or our finger. They both exist, but are so irrelevant kinetically that there is no interference. Thus they seem to each other to not exist to all possible measurements of certain resolution. But refine and increase the resolution enough, even with cryogenically cooled sensors, then that which is considered at first hand 'not', now has a glimpse 'being'.

That which is but a whisper between the whispers, the silence between the silences and the shadow between the shadows, can now be observed and measured.

This 'Wonderland', this Hypergeometric part of our superuniverse, is not so far away. Kaylie would find this out, and her older sister already had. For Kaylie humanly articulated, and for Alycia it was a curvature anomaly in a very small place in our local universe. Both girls would experience in time, the when or where that time is predominate in space, the here and now are subdued, proportionately.

It would be so relevant to what the girls and the old man had shared, when extrapolated for humanity to travel at very vast distances in space; it has to be as an axiom, that when something is, is far more important than where it is.

Kaylie had played with the oscillator that Grandpa had given her, but attached it on back of her mirror in her room. Pipa had shown her that oscillating the mirror would affect the light reflected back from it, especially at far higher frequencies. The oscillator was mechanical and so had put the mirror into a non-linear mode. The energy for the oscillator was of a very micro-sized nuclear source providing the electrical energy for the device. At the certain frequency, near what a Hypersphere would non-linearly function, Kaylie had followed Grandpa in putting his hand through the mirror. Yes, actually passing through it.

At certain frequencies, his hand would pass through the mirror and to the other side of it. It was as though the central part of the mirror was a quiet and subtle place of the local Event Horizon. The mirror in such non-linear oscillation at such a high set of frequencies is entangled, coherent to his arm yet in negative time partially, then in positive time partially. His hand remained visible as he would take it into and out of the mirror.

When his hand passed through the mirror to appear on the other side, it was still in positive time on this side of the local Event Horizon. It does not have to be a mirror, but a mirror is quite dramatic. Even a wall or closed door in such non-linear oscillations is passable. For at coherence, entanglement ensues for the mirror. It is the mirror's most central part, as for any similar oscillating plane that is passing through the local time and place of the grand superuniverse's Event Horizon.

Such novelty with the Hypergeometric machines was far more than just entertaining children. It was one of many astounding consequences of local and common articulation of matter and its very astounding reactions with the local Event Horizon.

One day Kaylie fell through her mirror. It was such that no injury occurred. Attach the oscillator to a wall or door of proper thinness especially and she could walk through it. The mirror, the wall or the door are themselves acting as the Hypersphere, but in a far more limited capacity.

One intrigue of all this for not only Grandpa, but his granddaughters as well, was the natural articulations already occurring in our natural world. Such proper oscillations in the brain of a child though seemingly rare, are the physical locations in time and space, between the sequencing of the flow of events in Hypergeometry and Euclidian geometry interactively of time and space.

So called by the mundane, 'flights of fancy', are so real to the participants in such glimpses of many wonderlands. Could the oscillating brain be a quantum machine as well? Then with such proper conditions, even the body?

Were they already surrounding the human theater of history's past incidences, and are there to be future occurrences? With the intimate connection between time and space, one affecting the other, has such been the unresolved norm just only glimpsed at so very rare of situations?

The farther things are away, then the more important of when they are than where. So to get to where they are takes far more understanding of when they are. Not just when, but the time and space geometry of when.

Grandpa had the Hyperplane experiment undergoing such greater endeavors than the Temporal Diffraction Grating. The Temporal Diffraction Grating was the quantifier of 'where is now', but the Hyperplane was the actual demonstrator of the conditions of articulating the local Event Horizon and the flow of data. This flow of data was then flow of events.

Events, whether linear or the more natural Non-linear, flow in the local direction of time as a sequence. Suppose 'zero through nine' in binary is repeated over and over, and so form an emitter, say a laser carrying such information to a detector; receiving such then, it is a repeat of the sequence 'zero through nine' in binary over and over.

With the Hyperplane, the rotating laser beam is carrying this repeated sequence. Place the detector at a radius where the beam's rotational velocity is below the speed of light, c, then sequence repeats at that detector the same. Place the detector at where this rotational velocity is equal to c, then the information is frozen in sequence. Beyond the radius of the speed of light, the information flow is reversed. At the radius equal to c, Grandpa called it a 'mirror'. This was the line of inversion between the Euclidian geometric forward sequence to the Hypergeometric reverse sequence of the same initial sequence. It was not an inertial Event Horizon so much, except with photons being massless, yet still left with momentum and kinetically able to interfere with the local here and now. But it was a border with each side replicating opposite flows of time or event flows of the same information, so like a mirror reversing an image, from the incidental to the reflected.

After such came the Hypersphere. This was the culmination of Grandpa's studies, being mechanistic and so much more than theory. After reading and talking to the old man it left me wondering less and less, and realizing more and more, that the machines were one and the same.

With all this powerful potential available, I had to ask him how far he had experimented. I wondered if there was even some nation, foreign or domestic that had tried to understand and even utilize such novel precepts and practices.

It would be another week before I was able to stop by again and visit with him. At our last visit he had given me a small disk to carry with me, less than one half inch thick with two small metal spheres connected by a spring. One side was plastic-like transparency and the other opaque, with no power source and not much to look at, but as a curiosity.

He told me to place it in my pocket or upon a table top. I kept it on my dining room table, and there it sat until I had another time to visit him. He sure was able to keep my curiosities going with all these new apparatuses, these 'Third Level Machines' as he called them.

The machines followed a formula that he referred to as 'The World Line or Point Formula'. The equation had an answer of ds^2. When the coefficient was positive, it was space-like, a 'first' level machine, such as cars, billiard balls and dogs. These we were very familiar with.

The 'second' level machines were quantified when the coefficient was zero. These were of only more recent understanding with the increased technology of the last ten or more years of radio waves and light waves. These are light-like and are only with momentum and no mass.

Then, with a negative coefficient are the Third Level Machines that are time-like, better time-predominate. These are the machines the old man was immersed in with his experimentations and theories. Hypergeometric Machines were all Third Level Machines. All the types of machines had elements of the other two, but each was predominate in either space, light or time.

The week was uneventful, actually very routine and there I was with a bit of a problem. I had an appointed office party to go to for the evening, so I just called the old man and let him know that I would be unable to stop by. I was tired and though he was fascinating, I had to have some kind of social life you know. I had hoped not to offend him and he was quite kind and wished me well.

"By the way, do you remember that disc that I gave you?" he asked.

"Oh yes. I still have it. It is quite an interesting thing. Sometime we shall chat about it and you can explain its special qualities to me I am sure," I tried to consolingly offer.

"Well, start carrying it with you in your pocket so you won't forget. Just everyday make sure you are carrying it, please," he intriguingly advised.

This disc that he had given me was another interesting machine. It was nothing more than a detector he had said. To me it was like an odd toy-like simple contraption. It was just a passive thing he had related.

"Oh my, I shall. I promise, and right now I am putting it in my trousers' side pocket and shall everyday carry it with me," I reassured him.

"Well, soon you shall see. For in a few days I will do one of my experiments, and you will notice the disc has use for something, for it is more than a little token of my eccentricities."

I would commit myself to keeping my promise and take it with me to the party and where ever else I went, trying to be as aware of it as I could without it being intrusive to my daily activities.

At the party I looked forward to the distractions and conversations of the merely mundane. I tried to keep in moderation my life with its boring to fascinating extremes. After a nice dinner and some cards, accompanied by soft background music that made a comfortable air about us all, I had my glass of merlot. Just as I was conversing I felt a slight movement in my right trouser pocket. It was a buzzing-like feeling. Then I remembered the disc and with amusement thought to myself that the old man may be up to something.

Upon the table before me I laid the disc down after the vibrations had subsided. The feeling of it was a bit uncomfortable and queer. I placed my wine beside it and as guests came by, they took notice of my gift from the old man.

"My, what is that … a pocket watch?" one asked. Others just looked at it curiously and continued with their courses amongst the party's guests.

Suddenly before me the disc moved to the vertical and rose and then hovered over the table. At first I was astonished, and in a few moments I grabbed it out of the air and tried to hide it upon my person. It was firm in its resistance, but once I had put it inside my suit coat it began to buzz or oscillate within itself. Then it tried to rise up out of my loosely closed side pocket. Up it came and I had to grab it once again. At present some people near me were quite interested in my toy and its behavior. Especially my trying to hide it after it had seemed to get away every so often from my possession, whether from a pocket, my hands or even after laying it upon the table.

I had had enough of trying to answer questions that seemed to imply that I was trying to be the 'life of the party' and entertaining everyone so well. I soon kept the disc within my pocket with my hand upon it, while sometimes it became settled or buzzed and began rising. I had to excuse myself, and soon found solace in leaving for 'another engagement' to afford my embarrassed state. My escape being slowed some by the continuing interests in my performance and my 'toy'.

Finally, having made my way from the party, surely to be talked about for some time, I headed home. I remembered that I had mentioned to the old man about my engagement for the evening, and felt that he had quite the mischievous side in his character. Along the way, there were a few more episodes of buzzing and then finally I found in my automobile some privacy and quiet.

Arriving home and remaining in the car, I let the disc just do its own thing and watched it as it hung in the air. I could spin it, but it seemed to reorient always to the vertical and when spinning would continue only if in the vertical. On its side, it would reset itself slowly to the vertical.

I now had something of substance to hold in my hand and to discuss with the old man. He surely had a fun time with me, as I had a struggle to keep his disc, my toy, from any more unusual comments and gawking from anyone around me.

It was later in the week that I had returned to the old man and gave back his disc. I had a smile on my face as I handed it to him and he was grinning too.

"Did you have a nice time at your party?" he asked, continuing with a slight smirk and placing the disc into a box in his bedroom.

"A grand time was had by all! I was very popular, and so much so that I had to escape," I responded with inflected humor, and knowing that he had me red-faced with his antics.

"Well, I did enjoy surprising you, yes I did. But it was quite scientific and demonstrating the wave nature of oscillating matter upon the Inertial Geometric, what others call the gravitational field for lack of better terminology," he replied as he offered me to sit with him again.

He looked at my bewildered face and then continued, "Well, you see when the Hypersphere oscillates, it actually sets up a wave pattern around it for quite a distance. Yes, it does. It is the disc's dimensions internally between the two masses that allow it to resonate in the wave action. So much so that it appears that the disc is hovering. It will tend to fit into its resonance well, so to speak."

I just smiled at the unusual old man as my thoughts focused upon the disc's functioning.

"It is a good demonstration aide for the common person. It is something one can hold in one's hand. It helps to make practical that which seems only theory," he explained.

"My dear friend, how far have you taken these experiments of yours? I get a strong impression that there is much more to reveal about what you have to show for all of your efforts, especially, considering the years of your work and life."

The elderly gent was very happy for my intensified interest in all that he had been sharing with me. Suddenly becoming pensive, he seemed to pause for a moment and looked rather sad. A longing came into his eyes as he became animated again, as if ready to divulge a secret.

"I would love to leave such things that you see here to my granddaughters. Perhaps that will be. But in the meantime, if something happens to me, would you please archive all these things? "

"I understand much of this has been sent to a special place already. Have you not alluded to such?" I returned.

"Oh my, years ago I had sent many papers to Israel. It has been the most powerful nation on Earth because of such technology. That was my very first move. In case of attack, Israel has access to such use of geometry in application. What shall the world do? Outlaw mathematics? That was my first assurance of the importance of this information. When all was done, I sent it to Israel," he said firmly.

"The potential for a weapon of war is there," he continued in a shaky voice, "but I would rather humanity reach for the stars and get to know the wonders of God far beyond their comfortable paradigms. God's universe, our local understanding of the universe and His Superuniverse ... we can only get a glimpse, let alone try to understand it. My, my, my", he murmured as he trailed off to sleep.

I was stunned to be considered to archive his personal grand work. And why was he so worried about his granddaughters' faithfully receiving all these things? He seemed heartbroken for some reason. He surely was a different sort, and others certainly loved his company. Perhaps in his family he had so long been misunderstood, and ostracized? Sometimes God has to give us from His family to replace those kept from us by our genetic relatives.

I could not heal the wounds he still felt, but I could rather be an asset to his work. It did fascinate me. I sat there as we both were quiet now. He was sleeping and all I could hear was an old clock ticking off the seconds.

Then I had a splendid thought. Perhaps he would consider my attempting to travel in the Hypersphere. It could be in reverse time or future time, all extremely telescoped. If his theories were valid, then I could for a short time, however that would be defined within all this new perspective, take a journey of remarkable adventure.

My contemplations were racing as I considered travel above the disc of our Milky Way Galaxy, and to utilize onboard instruments to measure temperature and radiation. To get images and powerful impressions of reaches into space that have been hitherto considered pure fancy, even by scientists.

Wonderland is possible, if I could be equivalent to a coherent particle, though I am macroscopic in scale, and in reverse time. I would need to be, by what I am trying to understand in his theories, entangled and so able to return to my origin.

As he snoozed, I stood from my chair and went over to the table with all his notes, papers and books. I flipped pages here and there, scanning with sharp focus all that he had implied. I looked back at his head upon his chest and listened to his snoring. He slept as a baby, but how far did he go? Now I was wanting to go farther and yet I wanted to be able to return. He was able to return, but how precarious was all of this?

If one loses their entanglement, they are then lost. It is not impossible, but it would take grand calculations to return if even all was functional and could be coordinated over such remarkable distances of reversed and forward time and space. After passing through the Event Horizon, it would be a strange wonderland. Even before that it would be a wonderland, but a bit more navigable, if one can define it thus.

If 'when' and time are predominate, then 'where' and space are less formidable. As he had mentioned about the tunnel diode and electron passing through an impenetrable barrier, so I would be able to pass through an impenetrable barrier. That barrier is space, great distances of space, but in reversed time. And with such distances, I understood that causality consequences were irrelevant.

If I were to go to a prior time, a distance of space would be required. One hundred years ago is about two and one half hours in reverse time, if the speed of Non-inertial Geometry is 370,000 times the speed of light. All that he had mentioned over the years and his writings were suddenly culminating in my understanding of what he had been trying to explain.

So one hundred years ago, is one hundred light years away, but is attainable in around two and one half hours in reverse time. That is the chord short-cutting the surface, the speed of light surface of our superuniverse, so fast that I would be traveling into the past.

That would be the easier trip for me to attempt. Farther distances like above the disc of our home Galaxy, would take longer in reversed time, perhaps days. I also had to consider if I lost my entanglement. If that were to happen, I could be lost in an alternate world line or alternate history, and lost above the Galactic plane. In an alternate world line of history, one would never be able to return, for the driver and source of the Hypergeometric Mechanics is in this world line. It is counterintutitive and that we just have to accept, I suppose.

I have so much to ponder and it is so very thrilling, yet so unknown. With distance being an ever present factor inherent to time, then paradoxes of causality seem to be quelled. If something happens to me and I die there, before I was born here, then here I just disappear, with no more trace of me since my departing through the Event Horizon. There, I just appear and leave any traces just after my arrival from the Event Horizon.

Such strange but workable stuff all of this, it was seeming to make some sense. It seemed to illuminate many things that had not been heretofore explainable. It is perhaps the Occam's Razor, the simplest explanation for that which seems to be difficult to conceptualize.

I left the old man to his solitude and headed for my apartment. Some serious planning would need to be done, just in case things would go amiss. I may be floating in the clouds right now, yet such things may actually be possible. I needed to get some rest myself, for there was much to prepare. If I could return that would be fine, but if not, then I had to consider that possibility very soberly. One must be able to accept the results of an earnest inquiry, for if not, then it is better not to inquire at all.

Chapter 8

Through the Event Horizon

Much of my pre-emptive preparations, mental and physical, were mingled with taking some quiet time to prepare for my journeys, for I hoped to take more than one. I needed the first one to just get the feel of all this. It was several days later that I met with the old man to plan an attempt. He said in order to make an initial flight for me, he would have to acquire more cryogenics, particularly liquid nitrogen.

Most experimenters used liquid helium, but all he required was the less cool liquid nitrogen. That was a great advantage for it was much easier to get this cryogenic and much cheaper. I'm sure his previous experiences had provided many ideas born from necessity. The old scientist was precise in his measures as well as his calculations.

When next I saw him, he looked almost young again. The anticipation had been good for him too. I know I felt like a 'kid on a swing'. We had much to discuss, but his mood was light and playful. "Do you think you are ready? Is it your time to go? Perhaps today is the day?" he queried with that old familiar Cheshire smile of his. And I thought to myself, 'I will remember this moment forever'.

He told me he had been doing maintenance on the Hypersphere rewiring parts of it, cleaning and modifying other parts. The oscillators and their magnetic amplifiers were tested. Monopolar vis a vis Farady and electrostatic generators, the capacitor bank, and other essential parts were tested separately. It was quite involved with computers that would control the signal generation for the Hypersphere and especially its point of 'slippage' as he called it. These were run by their programs and all this into dummy loads, then stop points in functions so as to test almost to an actual launch.

This point of slippage was the moment when in oscillation the Lorentzian effects compounded and a saturation occurred, and the sudden loss of mass and inertia. It is where length and time seem to shrink suddenly and the Hypersphere, as such a very slight disc to the eye, fades from view in a reddening of its light. In this state it is so near and appropriately thin compared to the local foliation of the fabric of our universe's space time, that the Hypersphere 'slips away', better tunnels.

Suddenly it is currently without inertia or mass to our side of the Event Horizon and as suddenly in reverse time, already to us at 370,000 times the speed of light. It has no more need for acceleration or deceleration, for it no longer has momentum.

Just as suddenly it is already racing at this speed on the other side of the Event Horizon. And with extreme reverse time dilation hurling in its short cut a geometric chord or tunnel, from our point of view through vast distances of space. This proceeds at nearly one light year for about every one and one half minutes on our side, thus minus one and one half minutes on the other side to us.

Each side sees its time flow as forward, as increasing entropy. But if one could see both sides at one time, each side is reverse in time direction or reverse entropy, to the other. Therefore the two cannot kinetically interfere.

So what we would see as tachyons on our side are actually objects from the other side traveling at 370,000 times the speed of light in reference to their side of the Event Horizon. Having the slightest kinetic interference and relevance to us, so the other side of an Event Horizon is the tachyon world to the side one is not only viewing from, but measuring from. These tachyons are worming their way through vast distances in space in reverse time. This geometric chord, this 'wormhole' or tunnel, is the subsurface short-cut for these faster than light particles.

Oh my, how I was understanding all of this so much more now. I was excited and overwhelmed by all of these conceptions and practices. Realizing the potential, I was like a teenager again, getting ready to take my test for a driver's permit, feeling as though I was ill-prepared.

It was with hesitation that the elderly scientist finally allowed me to go where his Hypersphere was located. I had thought it was somewhere in his apartment or in his neighborhood. But no, it was miles from our small town and at an old farm. He had owned the property for years and it was where he had done his experiments in seclusion, I finally discovered.

We pulled up and entered the old barn. It looked decrepit and abandoned to anyone passing by, particularly for scientific research. Inside it was so dark and musty, but there in the dim light I could barely make out some object.

The old gent turned on the lights and there it was, the Hypersphere! He had said that he usually kept it under plastic tarpaulin and hidden by sliding walls as well. But having worked on it lately, he had it somewhat more presentable for me.

I could hear an ever so slight hum and noticed that the Hypersphere was hovering above its driver ring a little more than twelve inches. It was nearly twelve feet in diameter and the ring was nearly one hundred fifty-eight feet in diameter.

Behind and around it were the support systems. The magnetic ring was the driver element as in a multi element antenna, where

here the Hypersphere, the mass hovering, was the actual 'signal'. It was similar to a normal radio antenna, but here matter was the signal to be transmitted or received, better yet entangled and tunneled through space, being articulated with more predominance of time and less for space.

Considering that with enough velocity, the measure of distance per approach or per recession involved changing the relative now between two objects. This is a derivative to time dilation. For example, just approaching one object's now is more future than an object remaining still. The same could be said for the opposite when the object is receding, the now for the moving object is the relative past to the still object. At the speed of around ten miles per hour of approach, an object five billion light years away is one hundred years future for its 'now' and if receding, then one hundred years in the past for a still object. If the approach or recession is perpendicular, then it becomes more parallel and the now is more the same, while the time dilation is the only remaining differential.

So this Non-simultanity is significant at great distances, depending upon recession or approach, even at leisurely speeds, when comparing where, or better when one's 'present' is to another.

If one can accept that no matter what velocity, even inconceivably slow, there is always some effective result of time dilation between this object and another object at relative rest. Even with approach and recession, the effect upon now is also true. There is a limit to practicality when these effects are in nano- or pico- seconds, or even at ranges of feet, yards or miles. They are minuscule, but they are still acting upon the relative objects involved with measurements between them. It is as though even in our decoherent world of the macroscopic, we have some background residual quantum-like effects just below the resolution of measure, but so impractical as to be irrelevant. Yet, as still true, we all are time travelers as well as space travelers. Travel in space, therefore travel in time; travel in time, and so travel in space.

All the Hypersphere is doing is amplifying what already is there. It is now geometry, no longer abstract or esoteric, but as real as mechanics can be. Its provocative results were effectively so predominate in time that far distant space was achievable. At 370,000 times the speed of light, entanglement and tunneling occurred for macroscopic objects. Thus precipitously did humankind light geometric fire that was dangerous as well as magnificent. When suddenly a macroscopic object is entangled and tunneling, it is coherent. It is thus able to leave the realm of being at our normal, more than one time at the same place, to be at more than one place at the same time. It is following that tachyon chord below the surface of the light surface of our local universe and the tardyon world above that surface. It is in reverse time, in contrast to the forward time above. For every year a light beam journeys, the Hypersphere also travels about one and one half minutes and so in reverse time. Now in six minutes the Hypersphere has acquired Alpha Centauri, and in around two and one half hours, it is one hundred light years away and one hundred years back in time.

This journey taking me above the disc of our Galaxy would require some days to achieve. I would have to keep entangled, or at the distances involved I could wind up a few thousand light years away and a few thousand years ago. I would be getting a great view yet be lost to perish long ago and very far away. If entanglement could be maintained and the tunnel kept accessible, then when I returned some days later, I would be comfortably ahead within the time I would soon experience, and back in synch with the world around me.

There was on board the cameras, Geiger counters and other quite elaborate instruments necessary to the function of the Hypersphere. Also personal items, adequate food and drink were secured. And I would be able to sit down in my modified chaise lounge or if need be strap myself to the sphere internally by feet spread upon the bottom and hands holding straps on the roof inside. I would see the disc of our Galaxy and its barred core as it materialized before me, as I materialized before it. In between I would not exist to the Galaxy, and the Galaxy would not exist to me. For I in this Hypersphere would traverse the geometric chord, the tunnel of entanglement, and so would be kinetically irrelevant. All around me would also be kinetically irrelevant. Wherever I had gone would now be kinetically relevant and would appear, as I understood from the old gentleman, as just some far away part of our local universe, only in reverse time. I would be amongst galaxies and their stars most likely in their normal time, but to me reverse time.

The cold liquid nitrogen coursed through the circulation system of the coolant coils for the magnetic drive discs. The subtle oscillations began, then more formidable were the increasing Lorentzian effects until the sphere hovered above the disc. Higher and sharper it could be heard and felt, the hum, throbbing and shaking of the ground. Even the air seemed to spin and breeze around, faster and more predominately did the sphere begin to squeeze vertically and redden. Now heat could be felt from the oblate sphere as it was radiating more intense energies of x-ray and gamma, reddening to a flat disc as it approached the Event Horizon. Here the star-like flash blindingly shook the barn and environs, then as suddenly it began to fade into the entangled tunnel. It was now only an after-image, for it was traveling at 370,000 times the speed of light and no longer kinetically relevant. The winds even blew the old man back and it swirled as a mini-hurricane strewing around bits and pieces of debris, paper, hay and anything not held down. Then there was quiet, and only a hum. I had left behind all I have known, and just trusting my old friend and his ideas, I was on my journey.

To me on board it seemed through the window that all the world and universe was reduced to a brilliant disc of light perpendicular to my axis of travel, and then that faded to darkness too. If one could imagine a red line tracing my course from origin upon Earth to a point about five hundred light years above the Galactic plane, one would notice the Sun regressing swiftly and the interstellar clouds and other stars racing by, as I coursed in representation from the laboratory in the barn to this elevation above the Galactic disc.

It was only when I had hit my programmed point of entanglement that I suddenly saw the disc of our Galaxy with its brilliant barred nucleus. Up to then all was dark and very cold. Only outside as it seemed to become stabilized did I again see a disc of light around my circumference, again perpendicular to my axis of travel.

I was there only for maybe twenty minutes, as suddenly I saw the universe and all around me reduced to a bright disc that slowly faded to darkness. A few days later, another brightening disc expanded to the view of the barn. I was so tired and overwhelmed with my experience, I could hardly stand as the door opened on the now floating sphere hovering upon the magnetic field. Then it just fitted down into its smaller hole that held it upon the magnetic disc.

I walked out on the magnetic disc's cover and to the ladder to the ground. With a hug and a handshake the old man beamed.

"Now you know! Now you know!" he said enthusiastically as he patted my back firmly. I could not respond or comment on my 'trip', for I was speechless.

We pensively shared a brandy, relaxed a while and then he contemplated,"Now you have traveled through the Event Horizon and know what I have been trying to express to others. You and my granddaughters are the only ones that know."

It took a few weeks for me to reorient to the normal world again. 'The implications of the application of theoretical geometry to practical mechanism are provocative', I thought to myself, repeating what he had told me so often.

Chapter 9

Into the Past

Winter was coming and the fall was refreshing after the hot summer. I had lost my inhibitions and now was of so much courage as to actually look forward to a journey of around one hundred years, and even meet my great-great grandfather. I would certainly be of discrete behavior so as not to disturb anything that was to come.

From 2010 AD to 1895 AD would be about one hundred fifteen years back and upon the Event Horizon of the superuniverse also one hundred fifteen light years. This was now almost second nature for me to consider. The formidable problem was to get back to my present here and now world line. And to do that I would have to make sure that the Hypersphere was entangled constantly. Otherwise I would be suddenly within an alternate history forever. I could not return, so as for all intents and purposes of 2010, I would be lost, never to be seen again. This is how causality is protected per world line.

I began in earnest to study that late nineteenth century era and the locale of my destination to Amsterdam, NY in the United States. At the time I had to consider that I would be a stranger to my ancestors who would not know me at all. Maye they would sense some resemblance, but they would in most cases never consider me as a descendant. They would possibly sense me as a distant relative though. I desired not to be an interference, but rather a very passive onlooker. My role as an observer was to make sure my visit would be of the least of kinetic inertial consequence.

My maternal great-great grandfather was Nicholas D. Simpson. He owned a building at 25-27 Market Street in this city and ran a confectionary with ice cream parlor. He baked cakes for birthdays and weddings, and was a fixture in that time and place for many years. My wandering into his store and then just back out or sitting there perhaps to sample one of his creations would be such a marvel for me. But would I have kinetically and inertially interfered too much? Even maintaining a secret knowledge of our relationship, I would still possibly be upsetting the course of events from that time to the future. Merely being noticed, even as a stranger, would possibly be interference. Perhaps for ethical reasons I would be better to dis-entangle the Hypersphere and allow myself to remain in 1895 and track another world line, so as to not unfairly affect the world line I had left. Should my personal, even selfish obsession be allowed to in time create such alterations to further events?

I longed to go back to the time they had lived. I wanted to smell, hear and see the smallest and perhaps insignificant details. It fascinated me so much. But I also did not want to have any divergent effects upon those things to come, whether to me would be pleasant or unpleasant.

I could already imagine the sights, sounds and smells of the time now so long gone. I would be an adult with mature abilities of observation witnessing my ancestors that would be impossible for me to do, if not for Hypergeometric Mechanics. To have just moments to live in that time, I would surely have some questions answered, perhaps opening new ones that I may never be able to have answered. For it seems when one gets answers for some things, more question tend to surface.

Another fascination in all this was that I also had a paternal great-great grandfather, Titus Cronkhite, who was alive at the time. Of course, the other maternal and paternal relatives of varying ages had to be round and about too. Still not allowing myself to try any kind of direct communication with any of them should be frustrating for sure. All information would have to be as passively received as possible.

My, if all this is of grand success, is it not so true what is said, 'we have a cloud of witnesses' and thus not only in space, but in time. So all that we do, even in private, is under witness, first by God and then in many cases by others we know not of.

I decided to not take along any papers indicating my identity either, for if something happened, then such could be, by information, an interference of the great course of events to come. As the days rolled by in my preparations, I mulled over and over such self-reflections.

To 1895 I would prepare to go, the same year that H. G. Wells published *The Time Machine*. Just the benchmark to honor this attempt of time, then space!

A fortnight had passed and it was pouring rain. It was dark since the sun had gone down and any activity within the old man's barn of the theoretical now mechanistic was quite unknown to the public.

"Accuracy is not exact," the old man had let me know, "and with that relative error so inherent, all we humans can do, observe and measure, is quite what will subjugate you my dear traveler to as precise a journey as possible."

With such relative error so prevalent always in human resolution, I was understanding from the old man that I could 'overshoot' and possibly be around 1860 or so. If for some unforeseen interference of the temporal kind, it could place me farther back, even to ancient Rome.

The general trajectory, not considering exactly all retracing in summation of the primary world line, would still place me toward the constellation of Orion. I was as ready as could be, for as an adventurer with great risk I had to be prepared for what could not presently be discerned with design. Therefore I left all details to the grace of God.

I headed towards the now humming Hypersphere, with its low frequency that one could feel. That, of course, was just the beginning, for when this geometry as mechanism becomes fully underway, its song of harmonics voluminously overwhelms with its far significant rise in pitch; also in its local effects that cause winds of cyclonic force to surround the environs of its operation. To so tighten local space-time curvature is also to unleash winds and shaking so prolific as to bring almost cataclysm upon anyone and anything within hundreds of yards of the Hypersphere's slipping away.

For me, within the womb of the sphere as it hovers and then tunnels, I would be impervious to such outer conflagration. But I would be in another dangerous situation where I might not be able to come back home, or perchance even survive.

I closed my eyes and let myself be within the hands of God, for what else could I do. In such a humble, vulnerable state, I was now at my most susceptible to forces and environments so very foreign to even the most learned of humankind.

Outside the old man could see the sphere just above the intense oscillating magnetic field of the driver ring. I began to compress within this ovoid that was reddening and flattening, and becoming more massive. Radiation levels were now increasing exponentially as the Hypersphere glowed so deeply red. Within I was beginning to see a ring form of the outside world. From the outside I would be seen along with this sphere of extreme geometry, as a flattened disc. Soon the foliation point would be reached and with that I would tunnel, and be entangled at 370,000 times the speed of light.

In slightly more than two and one half hours I was already at one hundred fifteen years ago, and one hundred fifteen light years away. With the consideration of hypergeometric mechanics I had approached our earth to around 1895 from 2010. And with amazing precision it placed me in the back of a building within a shed. Inside this shed, the Hypersphere materialized, and the wind and shaking moved the objects within this construct around. I had arrived on a rainy night right behind my great-great grandfather's three story building. The Hypersphere, kept entangled to 2010 and one hundred fifteen light years away, had delivered me with such astounding accuracy, that I didn't think the old man would believe me.

I quickly dressed in nineteenth century garb and made sure that I did not have upon my person anything that would create questions of any anachronism or interference from my visitation. I would get some sleep in the sphere for it was warmer than the inside of the shed. I tried to surmise what season it was from what my instrumentation was indicating. It was amazing to spatially be where I was, but I would have to insure that I was actually in 1895 or there about. I had to appear to be aware of the season, the relevant current events and situations of this place in time and space as to as totally blend in.

It had to be near dawn as light was appearing in the eastern sky. To familiarize myself, I had studied maps of the time so I could discern directions and thus local time. I looked around in the shed and in the increasing light I could make out a calendar; and yes it was 1895, the month of April. Of course of the thirty days there upon, I would have to mill around and gather information to know the exact date and the day of the week.

My clothing was not too new and yet agreeable enough to conduct myself upon the streets, enter into stores, so as not to be considered some derelict or vagabond stranger. I would have to take much precaution to try to keep the sphere and myself discreet so I could return. But how would I handle this Hypersphere in the shed, while entering and leaving for the next few hours or days?

My grandfather Nicholas' building in front of the shed had a bank on the right side, looking at the front of the structure, and on the left was his bakery and confectionary. There was an alleyway littered with debris and trash more abundant than ideal nostalgic flights of fancy would consider. Dogs were barking a couple of streets away and as the dawn welcomed the sun, roosters began their morning calls.

It was damp and cool, and I was getting hungry. I had made some modern copies of money for the time and hoped it would pass well enough so as to not be suspected counterfeit. I wanted to trade some of it and use other of it as to gain the real currency of the time. I was a bit tired, but would first have to consider where to stay, possibly to rent a room, yet close enough to be able to return to the shed. I hoped so much that no one would enter the shed and suddenly be alarmed by the Hypersphere. I covered it up with some of the boards and sheets of cloth to hide it. Already, the situation required quite a difficult amount of strategy to keep things subdued.

I discreetly opened the shed door and slipped down the alleyway to the street. I was so overwhelmed at my surroundings as to take every little detail into absorbed focus. The sounds, the smells, the details of little things like bricks in the wall and woodwork of doorways all were incredible to behold. I could touch them in their proper time. I was the 'stranger in a strange land' yet there was a familiarity.

The Spanish American War would take place in three years and I could not allow myself to dare mention anything of my historic future knowledge. I had to appear as normal of local time and place as those around me.

I dared to wander to the street by this alley, which would be Market Street. I remembered from old photographs that there would be businesses along this street. I would walk as inconspicuously as possible and turn left at the intersection to Main Street. There I would stroll along the streets as if simply another person of the time in these morning hours.

People must be going to work in the mills and shops, while others were going to breakfast in some of the local restaurants. I was amazed at the smells, even manure in the streets from the horses which pulled the wagons, carts and carriages. I noticed some telephone wires in a few of the finer offices and businesses, and in the homes of the more well-to-do.

Newspapers near the train station revealed the day and date. I noticed the calendar in the train station which confirmed the newspapers. I pretended to be waiting for the arrival of someone so to mix well and observe mannerisms, slang and chat. How they were speaking was as important as what they were talking about.

A train was coming as announced by a whistle and soon the floor in the station was trembling. I was seeing and hearing things from April 12, 1895 at 8:15 am. Birds flew by, dogs barked and a cat meowed, then screeched. Off it took as I just caught a glimpse of a dog chase after it and both were quickly gone. Everyday life, from the future so long past, now for me alive and real. I looked down and my shoes were dusty, and just under my left shoe, a fallen leaf from the prior autumn laid. These were common denominators of any time having the same conditions enough to be similar, not quite congruent.

I imagined if one was orbiting the earth and looked down it would not look any different. It was with more precise and myopic observation that the differences would show. These human-made differences would be the indicators of one's temporal placement in human history. In the natural world changes would occur in tens of thousands or more years. But human endeavors with fashions and technology helped to refine such indications far more frequently in time.

I acted out my character, left the railroad station and returned on the opposite side of the street on my daring path. I reached the intersection of Main and Market, and this time turned right and slowly walked up the street. I was approaching my great-great-grandfather Simpson's building and store front soon. So I stopped and soaked in the moment of this perspective and leaned against a telephone pole nearby, just watching and listening.

Some people passed by and I had already taken on the mannerism of touching the tip of my bowler hat, as a 'greeting of the day'. I would smile and say, 'Good morning' and others would return the same. There seemed to be no suspicion about me that I was aware of. I thought for a moment and tried to think of how I was appearing to them. Did I seem to have too much time on my hands? Was I out of sort to the norms of the area's expected behavior? I needed a newspaper to be reading and maybe have a cup of coffee, while I watched and listened.

I must have a story if I was approached. Perhaps along with waiting for someone, I could order a birthday cake at my great-great grandfather's confectionary for that loved one. I dared not get too complacent and blurt our something or act uncommonly as to have anyone wonder about me. I had to keep close to the Hypersphere too.

The Hypersphere, in maintaining an entangled state, was my only recourse back to my home and time. If the sphere itself had its own capability of providing the immense power needed to time dilate into the future by the generation of extreme Lorentzian effects, then I could possibly return. For if the shed is later removed or another structure is built there while I am in such an extreme state of physics, then I could wind up arriving inside someone's wall in the twenty-first century; or if a power failure, then in twentieth century or whenever! In this entangled state I do not have to time dilate in this space for I am already entangled to 2010. Thus a much simpler and less kinetically interfering proficiency exists.

A clumsy or awkward occurrence was much to be avoided, scientifically or socially. I found a newspaper laying on the street in good condition and picked it up. It was a few days old, so it had not rained if it was still there in the open. I could pretend to be looking for a job and carrying a newspaper would be a great way to continue this attempt at blending.

Leaving my musings I slowly turned and looked into the front of the building. Inside the bank I saw the teller windows on the left, with an office towards the back directly in from the door. I looked at a clock inside and then at my watch on a chain, which I had already procured in the future to help me blend.

I then slowly stepped back and shifted to my left to look through my great-great grandfather's store window. I pretended to focus upon the cakes and bakery items on the counter and behind. I tried not to stare inside at anyone in particular, so as not to stir any suspicions. A woman was hurrying behind the counter, probably my great -great Aunt Alice. I had to try so hard not to gape, but to put to remembrance as much as I could absorb about my ancestors before me. I was of the advantage of knowing some of them, while I was unknown of the future to them.

I deliberately opened the door and with careful bravado approached the counter. "May I help you, sir?" asked my Aunt Alice, while looking me in the eye with a gentleness in her countenance.

"Well … I was interested in a birthday cake for a loved one. What would be involved with its baking and icing and such? When would I be able to pick it up?" bravely I inquired with immense intrigue.

As she went into the back, my grandfather of three generations before my birth made his way out with supplies and busied himself behind the counter.

"Sir, may I help you?" he asked, with such a voice and familiarity, striking me inside my soul.

I could not speak for a moment, but yielded, "Yes ... well, the lady behind the counter had just gone to the back there and was preparing me information about a birthday cake." I was smiling at him, and he seemed to smile at me from deep within.

He went to the back of his store as she came out with the price and particulars for a two-layer cake, to be picked up prior to the birthday. Upon order, she advised that a day would suffice for its preparation and for my acquisition. She smiled and was very sweet. I felt familiarity with her too, but I dared not allow myself to obviously stare or seem to oddly linger in the store. I was anachronistic, and thus a temporal stranger. Overwhelmed I had to rush from the store, having had the semblance to bid a quick adieu and away I went.

Aunt Alice opened the door and called to me, "Sir, when do you need us to make your cake and when would be convenient for you to pick it up?"

"I need to find out myself, but thank you. I shall try to return in a couple of days," I returned, while not even sure I would still be in this time or for how long. I shuddered to have actually met and chatted with my ancestors in such an abnormal consequence of technology. They were in their late thirties or early forties, and I was not born yet. But, with Hypergeometric Mechanics, I was spending time with those I should not normally have conversation with, and sharing their series of 'nows' that they had lived.

They were my ancestors, by my time dead nearly one hundred years, yet now so alive. Making a living, they were making ends meet. He was sweating some, for he had just unloaded a wagon in the back for their supplies. She, with some flour on her apron, would only be known by my mother as a 'regal lady', so well dressed and owning a very polished piano that no one was to touch. My mother was a little girl when she knew her great Aunt Alice later in her life. I left the street speechless as I saw the 'past' all around me.

I had to maintain such a precise act to not cause any interference. I gathered myself enough together and slowly meandered back to the rear shed and the Hypersphere. Part of me wanted to just get out of this time and back to the twenty-first century, and another part of me wanted to savor 1895 and my ancestors, as they were.

In my great-great grandfather and even in great-great Aunt Alice, I could see the characteristics that were to be passed down to their descendants which would be found in my aunts, uncles and other relatives, including my mother, and even myself.

This was a very personal journey with extreme mathematics and science provoking thrilling personal consequence. Physics had taken to it limits and more, allowing me to be interacting with my so long gone ancestors. I had so many questions, but I could not ask them, at least not in any way to create confusion or suspicion. I could ask as a third party and pretend to be a 'friend' of someone that I was really related to during this 'present' time.

I was becoming emotionally weary with all this exhausting theater, and so I chose to retire to the Hypersphere as the day waned and dusk was upon Amsterdam, NY.

I approached the shed in twilight and ever so carefully made sure that I was not seen or heard. To the door and inside I slipped, not too dark with a little light entering through the dusty and cobwebbed window, I went over to the precious Hypersphere. It was warm as it continued to maintain an oscillation as it was entangled to 2010. There in its chamber was a flashlight. It was an LED type and small. I would be able to use it if needed and yet try to keep covert in my actions in 1895.

Night was falling and a cat was meowing someplace outside. I noticed a field mouse run past and a spider on a web in this small abode of utility I was sharing with them. With so much on my mind, my eyes fluttered and I fell asleep.

As the sun shined its morning rays through the window, I awoke to near silence, except for the birds singing the beginning of the day. I just laid there in the corner to rest in lethargy for a time. I was feeling quite relaxed when from across the street I could hear the sharp ring of steel upon steel, the sound of a hammer upon an anvil.

Chapter 10

The Past as Temporary Present

For me the last twenty-four hours had been Friday, April 12, 1895, and so it continues until I re-enter the Hypersphere and return to 2010. I was wondering if anything I had done could in anyway be of casual problems when I return. The Hypersphere had not disappeared and I was still able to do daily tasks, though not in my proper time.

The loud banging from the blacksmith's shop made me thankful that it was across the street. It was also the only sound of human technology I could hear, except for the more quiet sounds of wagon wheels upon the stones and bricks of the surface of the street. There were no telephones ringing for me to hear, though they exist for those who could afford them. If Simpson's Confectionary has one, I did not hear it. There were no radios playing, no televisions. I had seen a fan in an upstairs window in an office a floor above the street on Main, but it was not spinning. Telegraph existed with an office on Main Street and at the railroad station. All the sights and sounds were of just the local daily life. It was newspapers that conveyed any information from the outside world on a regular basis. Conversations of those discussing recent events in the nearby towns are the second best source of external information to those of this time and place.

The Sun was shining more strongly and brightly, so it should be a nice day. But the weather could suddenly change. And without our time's radar and meteorological forecasts, the intervening winds and such later today might involve a storm of grand proportions without warning.

With some semblance of courage, I again ventured out from this safe enclave of an old shed. But this time I would track down the alley again and bear to the right, getting another perspective as a stranger in a strange land having elements of the familiar.

Down the alley I carefully made my way. There was debris that I had not encountered yesterday, accompanied by flies. Someone from above in one of the apartments had dumped their chamber pot into the alleyway. It did smell some, but I tried to keep to one side of it. Old boards, newspapers and broken glass were still there from the day before. A wooden barrel with broken slats and rusted rings was also there. On this brighter day, I could see what looked like flour or sand of a tan hard consistency in the barrel's bottom and it seemed it had been there for a long time.

On the street once more, I made sure I picked up a strewn newspaper for I had forgotten the other one. It gave me something to fumble with and occupy myself. This charade allowed me time to think about my next move. I was not always sure where to go or what to do, for though it was the past, it still had possibilities of startling occurrences that I would be unprepared for.

What if I had an accident or a police officer stopped me? What if I got mugged or robbed? How would I handle a group of people finding me so suspect that it becomes a serious problem for my own extrication?

The Sun had risen in the sky and it must have been about ten o'clock or so in the morning. I had forgotten my pocket watch in the shed. Rather than wander back tediously for it, I decided to continue on up Market Street which was the beginning of a rise to a hill. I crossed the next perpendicular street and found to my right a stone brick structure. There were men loading wagons, burley and dirty with sweat. They had not shaved and probably were not as clean as the average twenty-first century laborer. I went on up the street, looking at my clothing as I walked. I was clean but a bit dusty, and I had not shaved either. I was fitting in well enough to those around me in this period of the latter nineteenth century.

In 1883 the super volcano Krakatoa had exploded in Indonesia, and I found an article about the event in one of the newspapers laying around. A team of scientists from The Netherlands, Great Britain and the United States were trying to surmise why this volcano had such abnormally immense power. A theory was already circulating, to be proven more evident in the future, that mantle plumes could be the source from the outer core of the Earth. Plate tectonics was still another theory slowly winning ground in the late nineteenth century's more universalist approach to geological understanding.

I could not dare say anything to concede that they were correct, for I am supposedly in 1895, not having such knowledge from the future. Another great area of speculation in the sciences was life on other planets. Mars was the big focus, and as by my real time in the late twenty-first century, much more had been learned of the red planet. Earlier life seemed to have been viable to fact, but at this time imagination and conjecture surely reigned.

Percival Lowell in Arizona and H. G. Wells of England would in a few years really get things going. Two years before the turn of the century, these ideas and the Spanish American War would accelerate the United States into an increasing race of technological as well as national progress.

I hopped on the electric trolley to Bowlers Brewery on the western part of Main Street a few miles away. Three cents was the fare, and luckily I had found some change on the sidewalk. I wanted to use as much of the real money, so as not to cause any suspicions about myself and my current use of this time. I considered getting a job, even if I were to be only here for a few days, so as to live more realistically like the people of this time lived. I stopped at a large brick structure of several floors and entered in through a wooden barn door that was partially opened. There was a man in the entrance, more rough than gentlemanly.

"Hey, you. What ya want here?" he shouted gruffly to me as I stepped in, still not adapted to the darker interior.

"I'm looking for work, sir," I humbly spoke as to represent the people of the time to those in authority.

"Well, ya look weak and frail, and I need waggoneers and haulers for these barrels," he spoke curtly. "And I have not time for your lack of hard work to be gettin' in our way. So go somewhere else!"

I left the premises, not wanting to get into any rougher a debate with him or the others working for him. They had all just stared, some looking as though waiting to be told to toss me across the road.

In short order I learned that most people were working ten hours days, six days a week for about two or three dollars a week. The workers paid for room and board or for a small two room apartment, weekly or monthly using most of what they had earned.

Besides the poor pay, the mills were hot, filthy and smelly. Most of the jobs there were not safe, with dangerous machines that caused eye or finger injuries or worse.

Not everyone was clean either, some able to bathe only once a week, so most had body and mouth odor, far more than I was accustomed to. Fingernails were often dirty, even in the Simpson Bakery. In small restaurants and lunchrooms for the working class even stained clothing of vests, trousers and dresses for the wait help were not up to our modern standards. The walls and tables needed painting or repairs and the windows were often dirty.

Without ammonite roads, the streets were always sharing their dust and dirt with those along their curbs.

Windows were opened to cross ventilate and with such opportunity flies were common in the lunchrooms. Often the cups, glasses and plates were dirty and stained with fingerprints. Street people were common such as vagrants and people with illnesses of body and mind. Oft times a policeman would make them move away from businesses' doorways even using a nightstick as a prod. No one questioned the apparent ignorance of simple health rules or maintenance of the street life at this time.

Out the front doorway of the upstairs apartments of the Simpson Building came a familiar looking man in his early thirties and quite dapper. He looked like the only photo I had ever seen of my great-grandfather, William H. Paton. He was holding the hand of the preadolescent George S. Paton, my grandfather. I could hear some of the conversation, and it seemed they had been visiting Aunt Alice.

In the west a thunderstorm was brewing and I could hear the distant booming. There was no weather service so everyone just seemed to prepare for the coming storm as best as they could. The breeze picked up and the dust spun around this way and that as some businesses closed their open windows. Another blacksmith kept at his business with not much concern. He was in his shop with the door opened and seemed to welcome the cooling breeze.

Stepping off the trolley, I walked east on Main Street toward the shed to check the Hypersphere. I did not want to just mill around on the street appearing aimless as that too would cause people to wonder more about me than need be. As I made my way up Market Street, I could hear a music box playing from within a small store's open door. I went in to get a closer look and found it was a 'Regina Model Number 13' playing 'My Olde Kentucky Home'. It was so simple yet so astounding to this time. The wind-up music box was considered by almost every patron within and passing by on the street, quite a marvel.

On my way to 27 Market Street by way of the alley to abscond to the back of the building where the Chuctanunda Creek could almost constantly be heard rushing, I passed by two older gentlemen.

I slowed up a bit to be privy to their conversation which had caught my interest. They were alluding to their view that General McClellan was the far more honorable general for the Union, than Grant. I pretended to be reading my worn newspaper that cost me nothing, but new was three cents.

The men mentioned that the local GAR chapter, the Grand Army of the Republic, needed a clean-up man. I was considering whether to stay a bit longer to try to learn more about these and other Civil War opinions from the mouths of those who had actually spent time in that conflict. This changed my plans quickly.

As I tarried, another man came out of the front door of the Simpson Building. He was one of the Ladd brothers who boarded there. He seemed to be going somewhere with a directness as if he had an appointment or maybe was going to work. I had remembered in my previous research that one of the Ladd boys was a bartender at the Breslin House Hotel on Washington Street.

Realizing I had strayed from course on the way to check on the Hypersphere, my interest in the war talk had side-lined me for a few moments. Believing that the Hypersphere was safe and sound, and with these new opportunities to experience everyday life of this time in the nineteenth century, I had to take the opportunity to subtly learn all I could.

It took a little time before we came to the Breslin House in the misty darkness and slowly approached the door to enter. The two older men never took a trolley, but walked briskly the whole way as

I just discreetly followed. I waited about ten minutes or so to make sure Mr. Ladd had gotten situated as bartender. Through the window I could make out a large painting of a boxer in larger than life display. I saw that is was captioned 'John L. Sullivan'.

I had my counterfeit money with me, so I could only hope that it was accurate enough to buy a beer. I planned to linger and listen, watch and live as life was being played out before me in this latter nineteenth century.

Inside the aroma of cigar smoke mingled with the odor of gas lighting, the burning of which produced a hissing sound as well as a pale low light. Its bar area and furnishings seemed old, even though I was sure it had been around long enough to be a respite for those needing to indulge their thirst and visit for a time.

Taking a bathroom break I noticed the long low porcelain trough and the wooden commode lined in porcelain, well stained from an overhead, gravity fed water system operated by a chain. I had seen all of these fixtures in books of days gone by, antique items.

I sat at a very small table to preserve my anonymity as a weary worker, and without light to read my newspaper, I could only order a beer. It was warm and cost five cents. I lingered and had a front row seat to life at six o'clock on this Saturday evening. Nothing major was happening. The Breslin was a benign place for the older gents to go in the evening after work and for some other men to escape from the house for a time. I was a bit noticeably younger looking than others of my fifty years. Health awareness and treatments of my modern time were like an aphrodisiac.

A piano player was lightly tinkling the keys for the guests of Breslin House as I sipped my brew. Some younger more roguish men sauntered in. Looking at me, then around the establishment, they must have surmised that none of us were a threat to their immature egos. They settled in quiet conversation at the bar.

Shortly thereafter I left the establishment and made my way along the dark street by gaslight back towards Market Street. I knew the way by daylight and did not want to seem confused if I wound up getting a little lost. It was so old-fashioned and tranquil, yet music could be heard coming from one of the other better bars along the way.

The walk once again brought smells and sounds of a time long gone ... meals being prepared in apartments and eateries, strange odors from the streets and factories, stray dogs barking and crickets creaking, and small talk from the mill workers that were finished. There was the usual, so far polite hubbub on the street.

A policeman reminding me of the Keystone Cops of old was strolling the sidewalk and playfully manipulating his nightstick. Farther up but on the other side of the street was another. Turning the corner of Main to Market heading north, I saw a group of young men gathered on the opposite side of the street just milling about quietly and staring at people. They looked at me as if sensing I was a stranger, but did not approach. Two more policemen were slowly patrolling both sides of the streets. I did not want to be involved in an incident that would call undo attention to me. Getting close to my alleyway I hurried a bit down to my precious shed.

I made my way safely, without notice into its confines, retreating into the darkness. I felt around a little and remembered seeing a candle and matches on a shelf. My LED flashlight still in my pocket, yet I seemed to prefer the old method. A welcome pale and dim intrusion through the window of a not too distant gaslight helped too. I lit the match and then the candle, setting it down carefully. Looking under the boards at what I surmised were rolls of dusty wallpaper and curtains, there was the Hyperspshere still vibrating and comfortably warm. Entanglement had been maintained, as I understood the mechanics of the Hypergeometric of our superuniverse.

I suddenly realized that someone else could be using this shed for quarters, possibly one of the vagrants. But all was quiet and looked undisturbed. Seeing a latch I set it to lock from inside the shed door and felt secure to retire for the night. I needed to wash and shave as well. I was getting to look more vagabond as the hours of time passed by. I would be just right to be the janitor at the GAR Hall. I wanted to listen in on the conversations of the Union veterans as they relived the Great War of the Rebellion. What was their point of view? What would I be able to hear from their own lips about those days that so strained Lincoln.

Morning came swiftly after a good night's rest and I was awakened by the Sun brightly lighting the shed from the eastern window. It was just a week after Easter Sunday and I could hear church bells all around Amsterdam, New York. Looking at my watch, it was half past nine o'clock.

I went into the Hypersphere to get my backpack with a shaver and some toilette products. I wanted to spruce up some, but still was limited to a wardrobe and range of travel. I had to constantly be aware of the delicate balance of my time here and all the requirements that would insure my safe return back to 2010 in the Hypersphere.

I planned to lay low today so I meandered on the streets making my way over to the GAR. I could sit there and offer my services for some pocket money as well as absorb all the conversation that the old soldiers would be sharing.

After early morning church services, many people were strolling in their best clothing, for it was the thing to do after church on Sundays. In front of the Market Street Presbyterian Church was a sign, 'Jesus is the Resurrection and the Life, No man cometh unto the Father, but by Me!'

I arrived at the GAR and noticed the door locked. Standing there for a moment, I pulled out my trusty old beaten up newspaper trying to look busy while others passed by. I had noticed people sitting on steps here and there in my short time of observation, and so I did the same. Maybe I would be able to be here when someone arrived to open for the afternoon meeting, if there was one. From my prior research I knew that they would gather at such places and even share a bottle of wine saved for the last surviving soldier of their regiment or unit.

In a shorter time than I had prepared to wait, two elderly men approached, slowly and haltingly for one more than the other. I offered my help at the door as they unlocked it and without a concern let me in. In my research I knew that in 1895 there were crimes, and Amsterdam statistically was within the average for its size and population of American cities of the time. I was surprised and pleased with the respect I witnessed on the street and had noticed the youth being polite, not insolent or argumentative. Though the men in their teens and early twenties would tease and taunt one another, it was not noticed by me of such towards the older generation at all.

Schools were six days a week, as was the work week. Then there were chores to be done around the house or farm, enough work for everyone, children, teenagers and young adults. It was rare for children to be so rebellious as to drift into the world of crime. The police, parents and even proprietors of local businesses were quick to circumvent a lazy or rebellious attitude in any subordinate, especially the young.

The freshness of all this, despite their hardships, touched me so that I thought not to return to the twenty-first century. It was like being on a tropical isle on holiday, when in the back of one's mind lurks the realization of returning to work and responsibilities. Oh, may I linger dear Lord for some more lost moments.

But I also recognized that I was in space one hundred fifteen light years away and one hundred fifteen years back in time from where I sensed I was 'supposed' to be. Yes, for all of God's designs are such, yet may we get but a glimpse of other times and spaces.

I asked if I could get a janitorial job for a short time. They seemed very accommodating. I told them that I was too young to have been in the war and they obviously could see that, but smiled at my naivety.

"Oh, young man, you may have not lived what we have, but you are welcome to listen to our old stories. Sadly, some have not been able to make a better life since. The injuries and the memories for some did not remain sweet or kind. All you can do is let them cry out with their pain of years of suffering and just listen. If you can do that you are so welcome to keep our home here, away from home, freshened up some." I could see in their eyes that they too had suffered unmentioned loss and pain. I made sure just to listen. But I did have a question that I kept for the proper time.

In my studies of the USS Monitor, I learned that McClellan was far favored over Grant by the officers and crew. Also there had been rumors, soon suppressed, that Secretary of War Stanton was really running things while Lincoln had been too trusting of others in his cabinet. What had they heard of such and even of the assassination of Abraham Lincoln? Intrigues I had but would wait to see if I could gather some insight from those who had lived in those earlier war times.

On the first of my work days, when only a few of the veterans showed up during the day, I had happily swept, dusted and set up the tables and chairs as needed. The resident bartender was rather quiet, but also kind and one of them. He was making pocket money, but for me this was my disguise. I had greater intuitions to learn from those who were there, who fought against slavery and a strong central government of federal over state power.

The potential for humanity was at the door at this time in history, my history, to take what science and technology were beckoning to the stars. If only the corruption and fools of distractions of greed and empire building could be kept in moderation. I was sad for I realized that such idealism was mute, for I could not or would not interfere. This history at the time of its making had to progress into what would come, as humans were allowed to govern humans.

As I had mentioned, I had come to 1895 in honor of Wells' *The Time Machine*'s publication. He would be later writing, and others too, of such concepts that would take humanity to the moon and their machines to the planets and almost to the stars. It was the ever increasing preoccupation of war and the destructive, greedy use of technology that would hinder more and more. Now in this time and into the early twentieth century we could still dream naively of human 'benevolent' and 'civilized' progress.

On Sundays only those who had to worked, and the stores were closed, so most people spent time with families. In the afternoons after church many strolled, couples and some with children scurrying about. I left the confines of the dark cave where the veterans lingered and stepped outside. It was nearing one o'clock in the afternoon. I was just enjoying the shadows of the north side of Main Street where the GAR was, and breathed in fresh air for a few moments

I thought I heard something, but passed it off as I could see nothing. It was difficult to identify what and where it was, but there it was again. It almost reminded me of a radio playing, but of course that was impossible.

"Mommy, I hope the police or fire bands play for us," said an excited little tyke to his mother as they passed by me.

Chapter 11

Precious Realizations of Yesterdays That in Tomorrows were Long Gone

I had so much food for thought already and more to come. I had 'visited' some of my relatives as real, living human beings struggling to make a living and in their own time, which I would not otherwise have normally experienced. I was able to hear those of a long time past expressing their views on happenings that would later be only re-writes of history from those who had never really seen or heard what they tried to describe. Now I could get unbiased opinions of such subjects that would be so politicized for modern consumption. Here was my grand opportunity. Here I was imbedded in the period of such old things that were, some that are, and others the beginning of things to come.

For many longed for the future, as many of us in the future long for the past. They naively believed that humankind would rise to some great pinnacle of development. Humanism was rife at this time and gaining with the advents of science and technology. Its philosophy of rationalism sweeping aside the semblance of conservative and ethical moderation was to them, a world of fantastic wonders awaiting.

The truth of a Creator-God was diminishing for many. Mechanistic was replacing the spiritual. It was a time when a surfacing logical understanding was replacing considerably a more supposedly 'irrelevant' divine.

Then I heard more harmonious sounds, now obviously the bands warming up, with the drums keeping a steady beat. The police and fire departments had bands that played in the parks for the general public. I rushed toward Market Street since that is where they seemed to be heading.

I understood that when the circus would come to town there would be a parade too. I needed to get a good spot for this parade even not knowing what the grand occasion was.

Usually, they were the only entertainment for the many, outside of saloons and theatre. I wanted to try to see some variety theatre, Vaudeville and now had a reason to stay a little longer. Next I needed to catch an early Vaudeville show. How could I leave already?

I overheard on my fast walk to get as close to this parade as I could, that a baseball game was to be held this afternoon too. What advantage for me to experience what by my normal time, was long lost and practically forgotten.

Approaching Market Street in front of the Simpson Building, the parade was manifest of song and visual excitement. Onward it came with dust rising around it in the street and blowing on the people; but no one cared, except for some elderly women in their finery. The children and especially teenagers climbed upon whatever they could to gain an even better view.

On the drums and horns played in parade march with glorious enunciation of brass trumpets, trombones, piccolos and cymbals. What happiness permeated the crowds as the wonderful parade went on and on.

There were policemen for control and keeping an eye on some seedier of the youth perhaps pickpockets or purse snatchers. Petty crime was one way to make ends meet then too, as well as another avenue of the perpetually defiant. Last week I had heard of someone being run over by a horse and another getting hit by a trolley. Apparently their foot was cut badly when it got stuck in the tracks, even though they got out of the way just in time.

I took the trolley to the park up near Rockton to the baseball game too, and that allowed me time to mingle a little longer with the people of this time.

It is written, there is a 'cloud of witnesses'; yes there is, in space as well as time, it seems. We never really know who is watching us or listening to us. It is also written, that sometimes we 'may be entertaining angels'. Though I do not claim to be angelic, it is interesting throughout human history that God has had His many ways of knowing human beings, even their most private manners.

If the old man from the twenty-first century had this capability, then as it is written: "there is nothing new under the sun"; have others in the past and also the future been of such to observe profoundly?

I arrived at the baseball game got off the breezy trolley with the sweaty and sometimes smelly crowd. I could hear the excitement of especially the men and boys as the game was already in progress.

This was the deadfall era and rules were slim. One could get hurt and still play until their injuries would sideline them. One young man had a hand of twisted fingers and was soaking it in a milky mixture in a wooden bucket. He had long ago been injured and yet I discovered he was a great player and a pitcher. He used to pitch the whole game and was readying to replace the pitcher on the mound. Some men were chatting that he would finish the game into gaslight.

When he smiled and looked at us, I noticed a few missing teeth and a bruise on the side of his face. This was the time that people were very adamant of the game. When something happened to either of the fans' disapproval, they would even fight in the stands and come onto the field where the police brought some order. From my research, the problem was worse when the police sided with the home team. Then the referees along with the rowdier fans were so biased it was a wonder a fair game could even be played.

This was tamer than what I had imagined that it could be. Not disappointed though, I took the trolley back to the GAR to finish my job for the day and perhaps to become more enlightened about the war, the men and the president.

After the parade it seemed more of the old men had gathered, I continued sweeping and listening. It was amazing to hear their stories from their true experiences. One gentleman had the scars from being slashed by a sword across the head and face, about three times that I could see. After some time, I waited to politely interject a question to these 'Boys of '61'.

"Gentlemen, I have heard and always wondered if McClellan was a greater general than Grant. I also wondered if Secretary of War Stanton was anyway involved in the Lincoln assassination. Could any of you give your thoughts upon such considerations?"

There was a moment of silence. I had then worried that I may have asked too poignant a question over the involvement of Stanton in Lincoln's assassination. Some of them may have devotedly served under him and even at this time, it was too fresh to consider possible conspiratorial theories.

"Young man, first McClellan, for us, was far more a general and gentleman. He was respected by General Lee and many of us also respected the great strategist of the South. We had often wondered along with the Monitor crew and others, why McClellan was suddenly not supported by Washington in the Peninsula Campaign."

My interest was so focused as I slowly took a nearby seat with broom still in my hand. I enjoyed listening to the war heroes as much as they enjoyed the camaraderie, sharing with each other stories from the past, as well as their daily activities.

"It was the Washington leadership that turned the military War Between the States to a political one. Many in the capitol, especially Stanton and Seward were running Lincoln's war efforts. Most of us fought for the Union and not for freeing the slaves. Many felt for the slaves, but our concern and devotion was for the preservation of the Union."

Another old soldier who used a cane had been sitting there nursing the same drink for most of the afternoon. I noticed that he had tried a few times to get a word in, and finally did, directed to me.

"Youngster, I find your questions very precocious. Yes, Stanton was power hungry, and many of us wondered if there was far more to President Lincoln's demise. Stanton it seems made sure he was not at Ford's Theatre that fateful evening. It always smelled of a greater subterfuge than the papers seemed directed to say. Even now in school the children get only an articulated story of those days and only from those publishers who seem to be the ones in sympathy to such ideologies."

Another gent raised his aging hand, trying to quell an uprising of feelings that could stir up a prolonged debate, "Gentlemen, gentlemen, may we now drink some toasts to our fallen brothers? First, for those of the Union and then another for those of the South who had only fought for what they believed was right. Then the last to the memory and honor of our great President Lincoln, whose grand face oversees our chambers of reflection here."

So well-spoken with dignity that one and all in his own way raised a glass of port, sherry and other wines or beer that had been now circulated amongst them all. He was a real peace-keeper.

Over the bar was a portrait of Abraham Lincoln and around the room some flowers were recently delivered by the widows and daughters of some of the soldiers. The room was not lit well by the gas lighting and even on this bright day, the large room's windows sparsely let in enough of the sunlight.

Everything was on the first floor on Main Street and upstairs I had surmised were apartments for some of them. Many came and went while I puttered around. I felt I may not make it as their janitor by spending more time studying them and listening and conversing with them. But that did not matter, for my main reasons for being here at this time were almost at a close.

I wanted to just observe my relatives as much as I could. Farther down Main Street to the east was where my paternal great-great-great grandfather Tunis Cronkhite was living. He was a stone-cutter as was his son, Robert Cronkhite, the father of my direct grandfather, Robert B Cronkhite.

Leaving the GAR I took a trolley to the eastern side of town. I slowly walked past 37 Church Street and looked in the windows of my maternal great-grandfather, William H. Paton's home. His son, my grandfather, George S. Paton was the child's voice I thought I could hear. I looked, but I could not see in far enough. A woman stared out at me from within, obviously perturbed by my peering into the window. She opened the door and called after me, "What do you want?" With a shrug and a wave over my shoulders I quickly moved on.

I know from a photograph that she was my great-grandmother Clara Simpson Paton. She looked regal and with bearing as my mother had told me. When my mother was little she remembered both her and Aunt Alice as quite sophisticated in dress and manner. William H. Paton was a high supervisor for Sanford and Sons, and so he must have been like that too, I assumed.

I was due to return to the Hypersphere, travel ahead through time and space, to arrive in 2010 AD whole. I had some concerns as the time passed here too, if all would be well for my return. So with sadness I had soberly prepared to leave from such a fantastic, personal journey. I was allowed to accomplish by God's grace, what normally is not supposed to be by human reckoning.

I had to really take seriously the return to my own time. The longer I lingered here in the late nineteenth century, I was adding more variables to the sequence of events from past to future. After I left and was in the early twenty-first century again, would there be any perceptible consequences from my being here? Even being as discreet and careful as possible, there must be some effects that would show, perhaps only in the most stringent of resolution kinetically.

As I slowly made my way to the alley and the shed, I saw Nicholas had the shed door open and was placing a barrel inside. It was of such a vulnerability I had allowed myself to be in. I certainly had enjoyed my day immensely, but I also had let my guard down some. So I went back up the alley to the street and waited just a bit so I would not be too conspicuous to anyone around my place of refuge.

As the neighborhood quieted more, I slipped back down the alley and made my way to the shed. Carefully I opened the door and relocked it within. This time with my LED flashlight I scanned around to see if anything was changed or worse revealed.

No, it was only for this new barrel being there that I had to squeeze around. I looked under the camouflaged covering of the Hypersphere, to see that my temporal and spatial conveyance was all right. It was still oscillating quietly. Maybe great-great-grandfather was a little deaf and not seeing well in the evening dim or maybe just so focused that he did not notice anything.

I looked inside the sphere, my only way to return to the normal time of my origin. All was well, and I needed to sleep. Tomorrow, would be Monday, April 15. I would only have to press a button and the Hypersphere would return in around two and one half hours to 2010. Entanglement was still in operation.

I really wanted to see the theatre hall on Main Street, to catch a glimpse of a variety show. Variety preceded vaudeville, which was growing because of the railroads. So I delayed my departure by one day to catch an evening set at the small Opera House near the Warner's Hotel there on Main and near Walnut. It was such an experience to witness the projection and elocution of the actors and performers along with the acoustics there. It was a little smelly inside and stained by the gaslights when they were high. But the show was a fantastic example of such entertainment to witness as it was then. All performers had been unknown to me, no real big stars. It was said the famous John L. Sullivan had appeared there. His fame from bars to opera houses was manifest again.

Afterward I made my way from the theatre and headed back cautiously to my shed. Perhaps on board the Hypersphere I could sleep a little. Once safely there, I was so exhausted that I slept on the floor. I could leave in the early morning, I supposed. Little did I know that I wasn't really wanting to leave just yet.

When I awoke, I glanced at the Daily Democrat, the Amsterdam newspaper. In it I read that on Church Street there was another group of Civil War veterans, the Sons of Veterans. As I walked along the streets I recognized those I would often see near, coming in and out of these two places in the city.

What a proud and sentimental lot they were. Already missing some of their comrades who had passed away with each year, but also those long gone, lost upon the hour of battle itself.

I found that more and more I was spending time mingling with them. I even took time to make some pocket money at the Sons of Veterans Hall, at Church, while staying on at the GAR Hall. I was managing to stay day by day and a week was almost at hand. I was acclimating to life in the late nineteenth century, that to return to my own time was sometimes bittersweet. Meeting my relatives was fascinating, but also the slow, laid back style of living, where each personal moment was so precious.

I was so immersed now that I felt I could stay here. If I were to stay, what would be the consequences? I knew from the old man that I, with the loss of entanglement, would lose my access to 2010. In that case I would disappear, since leaving my 'normal' time in the twenty-first century to never return. While in this 'new' 2010 to come, I would probably perish at some point before I was even born in the twentieth century. There quite possibly would be a grave saying I died in the 1900s.

It was so counterintuitive while considering alternate histories as such. The real historical account for anyone is when the kinetic signal to noise ratio is high enough to kinetically allow interface between anything and anyone. Occupying such as to form reality as we define the 'present' or 'now', we understand it upon the world line we perceive that we exist on.

I even thought as a test, to leave something behind and see if it would be in the normal 2010 that I had come from. It would have to be something that would endure the years and in such a secure place as to not in some way become lost due to any interference during its passage at one second per second into the future.

Perhaps I could leave a few things and see what would remain, just in case one or even two were lost. Thus, I would hide three one dollar coins, each wrapped in waxed paper and concealed separately. In the future years of 1960s to come, Urban Renewal programs of the Federal government would destroy much of what

Amsterdam was historically. So I privately knew where not to place things. This loss would include the Simpson Building sadly as well.

Where would I hide these coins in different places? It would be so wonderful to someday be able to check and see the results. If near a tree, it could be cut down; perhaps somewhere near the river or along the Chuctanunda Creek. But I knew there would be the razing and construction of buildings yet to come and what work in that creek bed I did not know.

I knew that the McClumpha Building on the corner of Main and Market would last until 2010. So possibly some bricks in the walls along those streets or in the alley behind would be wise hiding places for my time-traveling artifacts.

Then it hit me, and it would give me privacy too ... the Greenhill Cemetery. It was an archive and would give me sanctuary to place things and retrieve them, and to appear as though I was just visiting the graves. I could somehow even take the time to place a hole in the ground near the stones and place in each one coin.

The gravestones I was thinking of were not yet laid, for the ancestors of mine, that I knew well where they were buried, were still alive. My thoughts returned to the buildings in town and I was reconsidering putting one in each of the three areas that I had been considering previously.

As a few more days passed and the weekend was upon me, I decided to sample more of the variety theatre. The moon had been showing a waning crescent the last couple of nights and it was already April 19.

On the marque of the Opera House were listed some short skits and comedy, along with musical interludes. No one sounded famous to me, none to have made the early silent films to come. I went over to the more complete list in the doorway and one name did make me smile … a comedian up from New York City, John Bunny. I had read about him and even had copies of some of his silent film work on DVDs back in the twenty-first century. I was so flabbergasted as to think I could actually witness a performance of his. He was playing a forlorn hobo who would be outwitting the upper-class types in a scene of that night's show. There was an afternoon show on Saturday and another evening show that day too. I would go and spend ten cents for a mid-value seat and planned to enjoy every minute of it.

The working class really loved these shows where their aristocratic overlords were humbled by any lower class wise guy. This generated the popularity of the masses, while the theatre owners profited. The same stratagem would hold for the success of vaudeville and later silent films. At this time even the common ones were very familiar with Shakespeare's works too. It was refreshing to the lack of such in 2010.

Chapter 12

Indirect Return to the
Life and Times of 2010 AD

The next morning I stretched, feeling refreshed. I opened the Hypersphere and entered, feeling the oscillations of entanglement that also produced some warmth. I flipped some switches and allowed for the sphere to begin to oscillate more and more until to an outsider it would appear flattened, as to me through the port hole the outside world now began to flatten. A deeper, flattened disk-like redness would be how the Hypersphere looked to the outside world. The outside world would appear as such to me, then from each perspective again the dimming at the approach of the Event Horizon.

A humming probably could be heard outside the shed, and if anyone was to inquire within, it would have been to them a powerful apparition, so difficult to identify. Thus, with such an articulation of Hypergeometric Mechanics any past witness of that time in the latter nineteenth century would surely have been dismissed as fancy. In this, what I had accomplished was protected by its own effects beyond the understanding of any witnesses.

The whirring continued and as I awoke it was with much dismay. For on my watch it was two and one half hours later. As I had moved too quickly within the sphere's confines, I realized that I had hit my head and must have fallen unconscious. I looked at the digital calendar, noticing that I had been in reverse duration for over two days! This meant for the outside world I had gone over two thousand light years, where the earth was around 310 AD!

I let the computers run for now. I found myself on a hill with stonework near me. I set the sphere to entanglement again and opened the door. Stepping out I was in a warmer climate and looking down I saw a city. It had columns and quite the grand architecture. The scenery appeared Mediterranean around me, so I went back in and tried to get spatial, as well as temporal, navigational information.

From what I could gather this was ancient Rome! I was wearing clothing from the late nineteenth century with twenty-first century technology, overlooking ancient Rome of around 310 AD!

I dared not to step out for I had no way of assimilating into this time and place as I was. After half an hour or so I did step out and looked around. There was no one to be seen, but I could hear the shouting of a crowd, which would be from the Coliseum or Circus Maximus, as I could ascertain.

I returned to within the Hypersphere and set it to again approach the Event Horizon, but also to allow it to move and skim over Rome. If I could just get a glimpse of such a vista, then I would just dimly fade to a deep red, flatten, and hopefully return to 2010!

As I peered out the porthole I saw from above the ancient Rome of history, it too slowly fading to deep red, flattening and soon gone. Then I had to wait almost three days internally to return. I ate some, slept a bit and could only wait.

On the other side of the Hypergeometric chord was the old man already having been attuned to all that I had been experiencing from his instrumentation. He too observed the subtle after-image slowly appearing from deep dark red and very flattened to more round, brightening and becoming more solid. It was the same for him, the Hypersphere which would have normally taken about two and one half hours, now would take almost three days to fully return.

Hypergeometric Mechanics, the application of such physics to actual mechanism had been achieved, at least for me. I also knew that the quaint old man had done this and more before. He had achieved much contentment and confirmation in his accomplishment.

It took me some days to re-familiarize myself to 2010 as I had to evaluate all my fantastic journey. After returning I was frail and my head had a cut that had scabbed over. I was weak too from not eating, so astounded by my journey that I had not had much of an appetite.

As a fortnight passed, I was much more at home again. It was then that I remembered my three one dollar coins that I had placed in their respective places to meet me again one hundred fifteen years later.

I made my way discreetly in town to where I had hidden the treasures from the nineteenth century. I did not want to appear as a thief or crazy as I was searching these areas. So I went to the cemetery first.

I could not forget the 'Boys of '61'. It was wonderful to share their insights from their own perspectives and to see their faces and eyes as they spoke. As I passed by the many Civil War graves I realized that they had been gone for over one hundred years as well. These veterans had lives, memories and dreams, and looked forward to the next day's events. But one by one, each had passed. As it is written: "Life is but a vapor." Despite all the advanced technology and its fantastic possibilities, we still are in debate over life, and where it starts and ends.

I remember my father slowly passing away, reaching out to those he could not see, and even sitting up in his bed. I have heard from others that their loved ones while dying did similarly. Where was I one hundred years before I was conceived and born, when even a year before fills us with grand mystery and awe? Where have those gone who were so alive and close to us when they were here? These things are of the realm of God. Having had such an experience as written herein would provoke anyone of any true sensitivity and thought to also meditate upon such things.

Thoroughly I searched in the area of the cemetery where I had hidden the coins, but could not find them. I hurried over to what used to be McClumpha's corner and while no one was looking, I felt along the stone work trying to find where that coin was. It also could not be found. Then I went down by the Chutanunda Creek wall and felt there where I had placed the last coin, but again to no avail. No evidence that I had moved any stones or made any attempt to place such a monetary time capsule there. No scratches and no adjustments that could have lasted one hundred fifteen years, for it was as though I had never gone back in time-space to visit.

I finally found my way back to the old man again a few days later. On my way I was reflecting and contemplating over the idea that we may be vehicles, 'tents' as it written, and our physical bodies are so temporary. Perhaps as we are dying here, if found in His will, we are awakening there to a glorious life far beyond what we know here. And if we are found empty within, without God, then we are alone and cast away as our choice forever. Perhaps we really wake up!

I felt contentment as I sensed God in all that was scientific and mathematical in poetic and symphonic fruition. The Creator-God had all in His care. Yes, I was at peace, yet amazed.

Arriving at the old man's home, I made my way to his little apartment. I knocked but there was no answer, as was often the case. I respectfully and slowly opened the door and made my way in. As usual he was napping and I thus slipped into a nearby stuffed chair and relished his subtle intellect and his fantastic talents of

Hypergeometric Mechanics. I had such a desperation within me to try to find out why the coins, and perhaps all of my activities were not showing causality consequences in the present from 1895.

He slowly roused and opened his eyes. I smiled and he smiled back. "You have a question of many questions?" he so quietly mused.

"Yes, that is a good way to put it."

"You wonder more now than you did before, and are still getting used to all that you have been through?"

"Yes, so true. I believe you well know what I am going through, as you have already been there, maybe more than I know," I suggested. In response, he just smiled more broadly and I allowed him a few moments since he seemed quite peaceful.

"What happened to my one dollar coins I had left to be here when I had gotten back? I thought they might make the time journey quite sufficiently."

He was quiet for a moment, and then said, "At the time you were entangled in 1895, the world line split. Your only hope to return was as long as entanglement remained, which it obviously did. The coins moved into an alternate world line and future. The moment you arrived, already the alternate course ensued. You got back here to 2010 by the skin of your teeth, and only by the grace of God!"

He continued thoughtfully, "The split of world lines occurs sometime around the time passage through the Event Horizon. So causality is very well protected by the required distance in space, and also by the splitting of world lines. Just accept it. The only thing that came back was the Hypersphere and its contents within. Rest in these things, for they are only but glimpses that we are allowed. We have been given privilege and responsibility."

I sat deeper into my soft chair and had nothing to say. He had said it all and well. He only complemented what I had suspected.

I could not help but close my eyes and drift away in my new memories of long ago … some boys rolling a barrel ring as they ran with stick, girls with their dolls of the day playing tea party in front of a residence. Some older boys with their bowler hats, black vests, white shirts and dark ties, were trying to look tough, shoving one another into the street off of the side walk. It sounded as if they were with the 'ABC Club' that had a shack or 'headquarters' behind some building on Market Street's west side. They had energy and hopes, dreams and bravado. The Spanish American War was coming and in about ten years World War I, the war to 'end all wars'.

These young men were on their way up the street from Main and started to calm some as they had respect for the tall, burly policeman just caught in their view. The streetwise police officer had been watching them long before they took notice of him.

They became almost angelic as they passed him, and as he watched them go by. Their interest suddenly turned to a young lady on the other side of Market Street. She started to primp some and tried to act more mature. Now with ties straightened and bowler hats ever so slightly tipped, the young men winked and smiled, as she smiled and then turned away. An older woman came by and led her down the street, scowling at the boys.

On went life day by day and full of tomorrows, yet for me as I reopened my eyes, I realized such vivid remembrances. They were all gone now. I sobered some and turned to my old friend. He was again napping with a slight smile himself. I think I could better understand his contentment. I again put my head back and just listened and let replay the visions of my thoughts and memories I had just acquired recently, but from over one hundred years ago.

In doorways to some of the saloons someone played a piano ragtime style, or I could hear a rudimentary Dixieland sound. From the opera houses, for there was another that I had found later, old jokes, more music and Shakespeare, fresh improvisation by people making due with whatever they had. Some had discussed of what the future might hold, as already technology was observed accelerating. From theories profound to everyday sweet moments of details, all was so real yet now gone.

It is fascinating to consider that causality is self-consistent per world line, thus alleviating paradox problems. Such illustration is far better achieved when viewing the grander to smallest scales of space-time. First replace the primary of space with time, a simple inversion, then consider time as the prime function and space reacting to it. For we usually think of space flight and then the time dilation as a reaction. Turned around, long distance spaceflight is inevitable with the primary action being with time. This perception is easier by using the scales from cosmic to macrocosmic then to quantum as a spectrum consideration. Self-consistent causality is protected by the distance in space as per the distance in time, and at each reverse intrusion in time as an automatic change of world line offering a continued function of consequence thus initiated by the entry of a future to past event.

Thus when I had appeared in 1895, already a new world line that would be consistent was propagated or branched to accommodate this. While at the same time the original world line was sustained. So when I looked for the three one dollar coins I had hidden, they were not to be found in my 2010 that I was entangled to. Now if I had lost my entanglement and again used the Hypersphere originating it forward in time journey to 2010, then I would have traveled the new world line consistent with my intrusion into 1895. So this we can call 1895 AD or prime to 2010 AD which is the new world line propagated. Self- consistency then would allow me to be able to find my three coins.

All this alludes to the concept that there is an infinite or a finite 'imaginary' number of parallel histories quantumly apart, yet cosmically consistent; our observable universe with the related macrocosmic riding along functionally with our superuniverse.

Smaller consistent parallel histories can exist separately, but equally valid in greater consistent group histories, also remaining valid. Like reverse entropy of small systems, such as gaining energy, within large entropic systems, like our universe cooling. So our superuniverse is the pretext for our observable universe and other parallel universes being able to have forward and reverse time functions.

The quantum world and possibly the cosmic world can have both forward and reverse time actions. From observation from our macrocosmic world, we can see no difference from a millisecond to a million years, whether forward or back in time. We also can see no difference from a micrometer to a million light years in space. It is observed as coherent. When we do only observe in our macroscopic world of inches to miles, we do see things following forward only time flows. This is the obvious and easiest to observe and measure in our decoherent world. Replace 'world' with scale and all begins to make more sense, even that which seems counterintuitive.

But even in our decoherent or normal scale, according to Maxwell's Equations reverse time propagation of advanced electromagnetic waves were also as valid, though with much less strength of signal to detect, as compared to the forward time or retarded electromagnetic waves generated.

It is within the realm of Hypergeometric Mechanics that such exploits are possible, but alas just an expensive economics to glimpse of the greater things God can allow those who listen to Him.

Children playing and dreaming of life to come, as though a vapor long disappeared from our present; but it is the same for us. For in one hundred fifteen years, as long as the Lord tarries, we too are ephemeral along with our egotistical, haughty plans broadcast so. How now all seems, as the old man often eluded, so temporary, yet very important for the sequences of events ordered.

As mechanism, so sequential, yet with limited willful variance of choice within the greater determinism, we are still desperately responsible, yet without control more widely in such a Godly scheme. We imagine that we are so much more important in our decision making and our free-will is that we like to play God, but are only inconsequential to God. We, throughout our rebellious history with God have continually attempted to delude our arrogance to such comfort, while only fooling ourselves. In the very limited theatre of humanity that it has occupied for millennia we allow ourselves such intoxication. But upon the greater scales of time and space this falls so far from truth that we reel away from such clarity and rather run to re-indulge our addiction of egocentric aphrodisiac.

Chapter 13

The Old Man was Beginning
To Go to the Ages

He seemed more and more weary, yet always with his Cheshire smile. He had explored much and journeyed far in time and space. He was always letting it be known that he had only been able to 'glimpse' all that God had wrought. That is all humanity could ever hope for and handle. It brought him so much peace to realize that humanity was only given enough free-will and domination as to render in the end, the ever required need for God's guidance and will.

He had related to me that 'the past can only be read, as the future can only be written'. This based on the forward progression of time, which was inertial and far more easily accessible for matter and ourselves and our world as we essentially know it. He told me he had thought of this analogy when he was rejected by his son and his daughter-in-law, and so heartbroken that he and his granddaughters were so heartlessly separated. He would try to send into the future his messages as he was only free through time over space. His son and his wife had made sure that he would be restricted from his granddaughter for at least six months. He was much too educated was one of their many false charges.

He had once told me that in one thousand years or as the Lord tarries, future generations far more educated will look back at the ancient United States as they look back at ancient Rome and the idolatry of ancient Israel, as they defied God. This, he had let me know only encouraged him to be braver and more wise in the broadcast by pen and paper, now computer. The over-reaction of his children and their abuse of him for trying to share his intelligent reflections with his granddaughters and others who viewed his website, brought him inconsolable grief. He considered to reason that his own 'holocaust', his own 'Caliglia', were his son and his wife and others who found cruelty their only rebuff to well validated observations of the human zoo in history past and future.

He would often observe and reflect the present conditions around him from the perspective of past and future from a century to a millennium of temporal distance. Much better the whole context of the circus from such measure, than often in the myopic mix so immediate.

Less and less are those who actually study history, prone to blindly repeat it. In reading the writings of Josephus, one also sees the replay of the good and the ugly, the honest morality and the debauchery of humankind's defiant self-government while avoiding God.

The old man seemed more tired as the days progressed from when I had first met him. He had been to wonderful and tragic extremes in his life. I had sensed in the last couple of weeks that he was apparently more exhausted than usual. But with all of my re-orienting to the twenty-first century, I had not been so conscious of his slightly more aged manner. I did not fear losing him soon, but as I considered him more, I became concerned.

Was I anxious for him or myself? Oh, I thoroughly enjoyed his association, adventures and wisdom. I admit from some level of selfishness within me, that I also was feeling lost as if I wanted to again journey to a time and space that fascinated me. I was feeling more confident despite the ever present dangers of such Hypergeometric voyages. Without him would I ever be able to undertake such thrilling explorations again?

I went back to the barn and slowly looked around as to remain discreet, then I entered into the 'garage' for the Hypersphere. Inside it was dark except for light filtering through small windows here and there and the spaces between some boards. Under the tarpaulin there it laid in rest. I did not want to disturb anything but just take it all in. The old man had garnered disinterest from the many governmental and scientific concerns he had outreached, so as to at first discourage his work.

Here and there I carefully looked over and gently picked up items of interest to examine. On one table were a couple of the palm-sized discs he used that oscillated inside passively to measure vibrations and varying curvature of local space-time when the Hypersphere was in operation. I took my time with great intention to appreciate what the old man had done. His great passion and intellect had been so disdained for decades that he went on his own to first put into theory, then into experiment what consistently fascinated him.

I looked around and let my mind slowly consider and reconsider. Yes, I could do this, I could allow myself another voyage. I felt sure since it had been just recent enough that I remembered so much of what was involved from the old man's side of the operation. It was even possible, according to his manuscripts that an automatic mode would do very well. I just prayed to God silently that the power system for the Hypersphere while I was using it to visit some time and place, long ago and far away, would not fail. Yes, there was a redundant back-up power system, but it had to fully operate without any loss of entanglement between then and now, there and here. Going into the future was not a problem, for if the power died, then the extreme dilation that I would be under would just bring me into a future time, but still here. Instead, coming back from a past time to now was the problem. For if I lost my entanglement, say in ancient Rome, then I would lose my present world-line. Yet from ancient Rome I would wind up by pure dilation of time at that future point to 2010 AD or prime. So I again would not need entanglement to the future time in the Rome of the future, but would be on the alternate line of causal history from when I first arrived in reverse time to the ancient Rome of whenever past. Thus, from the past to that future, obviously no entanglement would be needed.

I went to my apartment and rested and could not stop part of me from dreaming of a return. It was addictive that to be able to actually visit historical times was now possible. I could not easily remain practical while desiring such adventure.

A few days hence I went to visit my old friend, but he was not there. I knocked at the door and he did not answer, and when I looked into the window all was seriously quiet and empty within. I felt unexpectedly sad and could not help but tear up. I already knew inside what had happened. I was suddenly now very alone and had no one with whom to share any of such grand schemes in Hypergeometry.

Shortly at my apartment an envelope was delivered and then a message on my phone notifying me of his passing, funeral and burial. It was from one of his granddaughters. The old clock in my living room was ticking so loudly but the powerful intermitting quiet could be felt. It was sunny this afternoon, but I just sat in the large easy chair in my living room. I did not want to move, but just sat quietly surveying the room. He had usually napped around now, but even then he was here and near. I shook my head to somehow sober myself. I really did not know what to do, but I did find solace in contemplations of the temporal kind.

Ancient Rome, 1895 AD, the future, the past and time as well as space was before me. What glimpses could I be blessed with? No matter how far I could travel in these cooperative regimes, I still would within myself, have maybe enough years left of my life to complement the basic human lifetime of around three score and ten. What would I do with the time I now inherently may be allowed and where and when would I spend it? I may have an accident of death that would so accordingly truncate my statistical allotment of common human years of life; but still I dared to reflect more intensely upon the potentials before me.

A liberation was within me as I still cautiously, yet surprisingly with some boldness, thought to actually try another voyage along the temporal road with the past preferred. Why the past? For the past I knew more certainly than the future. Certainty gave some comfort in familiarity as I have already eluded to. Perhaps the future, but how far into the future? Any journey either way had increasing risk in case of loss of entanglement. In the past I knew that I could still enjoy a benevolent environment and climate and would be more aware of the surrounding social instabilities as they may impinge upon me personally.

Even with such advanced technology I still had to consider my own survival. There was always in the adventurer's situation the element of risk, no matter how safe one tries to attain such security.

Ancient Rome in the early fourth century was the trip I considered. There were so many others as well. But to just ponder a return to this Rome of history was to also consider many other things as well.

Disease, so I would need to bring some antibiotics. Injuries would require some kind of first aid equipment along with emergency food and water. This was also a concern in the nineteenth century as well, but the old man was so thorough that I had not thought of it. Now I wondered if he had.

The mode of dress for the late nineteenth century had not been too difficult, of course I had to avoid modern materials. But for ancient Rome, I did not have the resources to really be sure. There were illustrations and even colored paintings since the Middle Ages, but how accurate were they? My quick 'accidental' sojourn had offered so very little as short was my time there. From a distance some appeared in rags, not the clean, colorful textiles of wear that many of us have seen in our modern, present time. It was quite the 'sticky wicket' to not be in anyway anachronistic for the natives there. I could not initiate any distraction that would arouse suspicion of my identity. Some might find it interestingly odd, but others could panic to the point that my safety would be compromised.

I had to still do more research and gather enough coordinates not only for when I would be traveling but also where. Such a distance of 1,690 light years in reverse time required precision indeed. From what I had gathered from the operations of my nineteenth century trip by the old man, there were slight stages of arrival. First the system checked itself by arriving just above the atmosphere where one could observe the earth below for some moments but in that time. Then would be the surface encounter and sensors would scan for any object so the Hypersphere would not become part of a solid object.

Next would be the arrival within an enclosure, like the shed in Amsterdam, so as to give some discretion for such an unusual apparition to any local, past time observers. This would at all times require a constant entanglement to be maintained as well. Also a scan would need to be made of the interior and surrounding outside area so as to insure insulation from kinetic interference, until I could personally make my own observations of the immediate surroundings.

This would all be required for fourth century, ancient Rome. The accuracy would be less for my blending in, for not many records were available to me. With no photographs from that time I could only imagine how the more common people dressed. It would be a far more difficult venture. I also did not speak Latin, so would have grand consternation to even pick up the local tongue around me at the time. I could be biting off far more than I could chew this time around.

I sat back in my chair and could not help but contemplate what so many fail to observe. Event sequences with culmination of deterministic fulfillment can take the easily accepted few days or perhaps even weeks, months or years; but decades, centuries, thousands or even millions and more year. How we humans love that which we consider reasonable and easily defended with the locality of here and now, not too far out of reach. We claim so often science and objectivity, while humanly limited to ideally attain. Mathematically such is a limit never to attain but only approach. Yet, even our most arrogant academic in their private ivory tower will disagree, while hypocritically proving such, at the same time.

I had learned from the old man not to waste my time with those who claim to see so well, while revealing how blind they are. For when he tried to share his discoveries with the scientific and government establishments of his earlier time, they essentially ignored him. While so disregarded he continued to move from his primary experiments to far more advanced ones, alone, but for the grace of God. This tenacity within him inspired me and the few others who cared to really listen.

I had to prepare quite definitely and with well repeated reviews. I could not engage this endeavor with any superficial efforts, for that could be more socially dangerous rather than technological, upon arrival and visitation. I had to ensure my survival despite the wonderful rapture of my being in the fourth century and to witness ancient Rome from common everyday ground; ground-truth to the extreme.

The day and hour had arrived for my journey to 320 AD and ancient Rome in all of its experiential of hour by hour and maybe day by day, if all would went well.

Inside the Hypersphere I had preset in redundancy the analysis program the old scientist had written. All I changed was to enter a time farther back. I made the necessary adjustments for the coordinates of the position of the Earth, Sun and our position in the Milky Way Galaxy and our Galaxy's traverse along their joint interactive world lines. I was not even sure that I really knew what I was doing, but for the confidence I had in his work. He was so successful with my nineteenth century trip, I was sure an extension of his algorithms would suffice for this adventure. If anything seemed wrong, I could send a reset as long as entanglement remained, and back home in time and space I would be. I just hoped that my amateurish efforts compared to the precision of that wise old gent I so missed. I hoped that I would fade into view within a safe spot so as to not shock those I would meet later on in the year 320.

Chapter 14

Unto Ancient Rome in 320 AD

After reviewing all of my complement of materials and needed paraphernalia, then rechecking the computers in the Hypersphere, critical of performance and precision for such this journey, I was ready to begin.

Once set, the computers, driving ring and its power sources of high current, magnetic fields' amplifications, oscillations and hypergeometric functions would be here and now on this side of the chord between this 'here' and 'now' and the here and now of 320 AD. It had to maintain the entanglement required for my return.

I set the countdown into the onboard computer which 'hand-shook' with the base and all sequences began slowly. This trip, it was up to me to make sure of everything, for the old man was now in a far better place.

Again, I could see through the portholes, which had been kept small to reduce incoming radiation, the outside world pinching into a compressed disc around the sphere. It thinned and brightened, then faded and reappeared later on after a day and a half or so. I tried to sleep or review things, but as on any exciting trip it was not so easily accomplished.

According to indicators and graph generators onboard I was nearly now at the goal of 1,690 light years as well as the same back in time. I was now able to see some of my approaching surroundings, but only within a widening disk uncompressing. I was so hoping that I was in a convenient space-time for I did want to be able to come and go and not surprise anyone nearby.

As all this settled to a consistent low subtle hum, acknowledging entanglement kinetically active, I waited for a completion enough for me to egress and explore. Soon the familiar silence except for the very hint of oscillation of entanglement reigned. I looked around and it appeared dark, but for some tangent, faint light from a single direction. I slowly opened the Hypersphere's only part that actually travels, the spherical part itself, and gazed a bit better around the dimly lit interior surrounding myself and the machine.

As my eyes acquired the precious light, it revealed a stone work around some kind of warehouse structure. At the far end was what looked like a doorway, which opened to a bright outside world of trees with leaves blowing on this sunny, breezy day somewhere in the fourth century, and hopefully in ancient Rome. I supposed that where I was standing would one day be part of dilapidated ruins under some layers of earth and quite possibly only by a long shot discovered.

I could hear water dripping and wondered if I was under an aqueduct and this was part of the base supports that could be also used for storage. I slowly slipped out of the sphere and approached the doorway. I stopped at it just inside to listen if there were any people outside.

Carefully I peeked around the support on either side and found that I was right about being in the base of an aqueduct. It was located in some kind of a grove lined with Italian Cypress trees, Cupressus Sempervirens that ran parallel along the length of the aqueduct for some distance. I stepped out all ready in the best of period attire I could muster from the twenty-first century to blend into the fourth.

I did not look too bad and felt I could fit in for a while. I could see movements down the hill between the trees and noticed some people. They looked somewhat more ragged and not very healthy. A few were lame, others were just sitting, some very old and a few children. There were barely audible conversations going on, but of course I would indeed struggle with that. I would have to play the part of a deaf-mute or maybe act lame. I did have some arthritis so I could emulate it far worse than it was.

I began to make my way down the road with trees on my left and the aqueduct on my right. The sun was to my right and the shadows indicated it was afternoon, so I was heading south.

It really was very quiet for I was not yet accustomed to the absence of motor sounds or radio or background music. Here there would be no such marvels for another 1,600 years or so. I would have only my hopefully well- hidden Hypersphere to remove me from this even more daring venture to return to the mid twenty-first century.

I walked about half a mile and looked between the trees to what looked like ancient Rome, but vibrant and not as sterile as a museum model would be. At this time it was still in its proper time and flow of days to come.

I slowly turned as a man on a horse approached, prepared to play my part well. He was a Roman centurion and his steed had a slight limp and battle scars on his body. Soon he closed in front of me. I just acted as though I was a bit abnormal and he seemed to perceive I was a misfit. He looked me over carefully as if I was a stranger worth investigating. Perhaps he thought I did not quite fit into any class of fourth century Romans.

I let down my head trying to hide that it was an effort for me to just stand there under his scrutiny. He swiftly unsheathed and placed his sword under my chin.

"Qui es?" he interrogated inquisitively, not threateningly. I was numb and had nothing I could think to say or do. I knew so little Latin, but understood that he was asking me who I was. He replaced his sword and said something to another centurion on an accompanying horse. Both of them looked at me and then spoke to each other laughing.

All I could think of was to sort of bow to them and act like I was somewhat queer. They then slowly moved on, hopefully convinced that I was of no threat. I had doubts of the wisdom of my coming to this time. In preparation I had superficially reviewed only enough Latin to get by. I had arrived with some beard stubble and it had grown more in the last few hours. But that was fine, for I had to allow myself to be dirty enough to fit in. Our twenty-first century hygiene was unimaginable in this fourth century since Christ's entry into recorded human history.

I also felt a slight irritation under my chin where despite the thickening beard hair, the centurion's sword was sharper than I had imagined. There were some blood spots on my fingers, as my hand would unconsciously go to that place to relieve the slight itch that accompanied the scratched area. It was as though some of the beard growth had actually protected my neck in that spot.

I had gone back to the hidden sphere to make sure I could find it quickly. It meant not wandering too far from it if I needed possible emergency tactics to escape this fourth century Roman locality. I slowly went out and down the path again much more wary and alert of my surroundings.

In this time the Emperor was Licinius, a persecutor of the Christians after a period of tolerance. I was able to observe much as I was making my way into the City of Rome along a path, which many considered a via or road. It was not that wide but did have stonework upon it for times. Other times it was simply grassy and dusty. I could see ruts here and there that would have been the width of chariots I surmised. It had to be very old, probably one or two hundred years, maybe more, but still being used. As Licinius was now the reigning Augustus of the eastern part of the Roman Empire, he was also reneging on his Joint Edict of Milan to allow and respect Christianity in the empire. He was a contemporary and a severe rival of Constantine I.

Supposedly, Licinius' wife was a Christian, but it was understood that he feigned it and his recorded cruelty seemed to indicate thus. He was the sovereign and perhaps the Roman soldiers were looking for Christians to hunt down. I may have not realized it important enough that I had more to bargain for than merely being a strange man out of my proper time.

I had studied more carefully than I had before my nineteenth century trip, the particular history and even the locale of my present venture in time, as well as space. But again, now kinetically able to interfere or to kinetically interact, I was not only facing wondrous opportunities to actually witness history, but also to be harmed by it.

I could hardly escape such thoughts of not only the scales of time, but of scales of space. Yet before me was great history and knowledge of the time, some no doubt later to be lost. Their ignorance of the inheritance they would be leaving their descendants that could have made life easier, instead of just survival at the beck and call of a far too pampered elite.

I had to sleep so I returned to the security of the machine at the base of the aqueduct. Here I had some control, while still wondering who or what might intrude during the night. Here no more historical theory of antiquity, as day to day and moment to moment I must be wary of life and limb. Survival was now becoming much more the priority than esoteric reflections of being safe with my warm Irish coffee by a nice fire in the fireplace. Now if entanglement would break down, then I would be forced to live here in the past and accept the severe results.

I actually missed the familiarity of 1895 and its recognizable characteristics that I had enjoyed. Here I felt more on the defense to survive and was concerned that such in situ studies could wind up tragic for me. Now as I was so kinetically interfaced to interfere in this time and place, it was presently my real 'here and now' that was the prime determinant of my life. Theory in practice was now essential machination. It was the same for the seemingly abstraction of geometry, now mechanism of this present, very real journey. Advanced scientific thinking came upon me so suddenly in such ancient surrounds that I realized how grand this venture of time was over space.

I had to remain in my sobriety to be as cunning as possible, for this fourth century was far more unpredictable for me than the nineteenth. Thus, as with increasing distance in space, the increasing distance in time was also proportionally problematic to the safety of such travelers.

Awaking refreshed, I looked around and found no indication of any intrusion into the hide-a-way for the Hypersphere and me. I cautiously exited, looked carefully around and admired the beautiful sunlight, thankful I had a moment to listen to the birds in the trees.

I went back into the larger stone room and slowly approached the door, but I blocked the light with my hand to make out anything amiss in the dimness. So far, so good and all was well. Reappearing outside I was aware of my surroundings and the scent of flowers and grasses in the air. I approached a spot between two of the Cypress trees and looked out at a fog lifting from the lower parts of Rome. Straining, I thought I could hear something, but it was so faint, I was not sure. The sounds were a cacophony so distant and vague of whatever could carry through the air to my attentive ears. I wanted to be safe, yet to discover. But I had to have courage, for if I ventured into some situation I could not extricate myself from, then I would be in quite formidable distress indeed.

My curiosity would not quell, so again I dared make my way down the simple road and allowed myself to venture farther, but with heightened caution. I could not stifle my desire to saturate myself within the year 320 as my here and now.

No one appeared to rush in these times of slower tempos of life. The pace of life seemed to follow the pace of general transport of technology. Transportation reflected this as well, first mostly feet, then horses for a few and as the ages passed, motorized vehicles. In more time the spacecraft and computer proved that we prefer to move at a speed which increases with the pace of life's duties.

I kept strolling along and noticed that I was coming to a small gate into the great city of Rome. Inside it was bustling with activity, markets and people hauling things and conversing more intensely.

It was late afternoon and I could see beyond the stone walls the buildings and columns of the time, which showed some stains from the centuries of rain, but also the grandeur I recognized in the ruins I had seen in paintings. But actually here, what a sight I was to behold. Along the way and at the base of the columns were the poor, the lame and crippled begging. Foul smells, dogs and chickens were mingling with the beggars on the byways. The elite Roman soldiers looked as though they were almost strutting while the lower classes scurried out of their way. Here and there a Roman soldier was keeping the peace.

Typical of humanity, for our basic nature has remained unchanged, history has so often repeated itself for the many who care not to study it for its revelations. Selfish interest and ignoring past history, they choose to lose their sobriety and restraint of knowledge and discipline. Then for those of us who love to study history, we see generations repeating the mistakes of those before. Wars continue, the poor starve and do not get their justice, and the cruelty of human upon human does not abate.

Our achievements in technology have only raised us to fall deeper into a more sophisticated, high technology, polished feudal system. One against the other, and the illusion of liberty allowed to program the masses on how to behave. Play the game, follow their rules, stay in your place and you will be allowed to survive. If you are in the elite of the said feudal states, corporate or national, then you can be above the consequences of laws forced upon the commoners. To question too much is treason.

We in the modern times have copied Rome's elegance with brutal control. This ancient Rome presently before me was as it really was, without cosmetics or objective. Now I found myself walking down the small street amongst the little markets all crammed together so closely. I tried to blend in, but occasionally caught the gaze of others who seemed to find me an odd one. I made sure I was dirty and bearded and even continued my act of having some issues of hearing and understanding. I had no choice. Now I was growing in my concerns that I had entered a world I was not properly prepared for.

I noticed the sky above me as evening was coming, accompanied by dark clouds. I had perhaps wandered too far to get back to my Hypersphere. I would indeed need to achieve such a common Roman life, for I may find myself sleeping in some quick shelter from the rain and possibly cool night.

Where would I go if I needed to get shelter and food? Without any Roman money I would have to possibly turn to petty thievery, but that could bring consequences that I may not be able to overcome.

Hypergeometric Mechanics, once for me so theoretical, was now proving so survivable in results. Beforehand I would romantically dream of such a venture, now I was almost regretting it.

Onward I ventured, but finding some discreet hiding place was next to impossible. I was getting nervous as I felt that some suspicious of me were following. I acted with a facade of confidence to non-verbally show myself as not a disturbed person, but one on an errand for a more important citizen. I could not think of anything else for I had to move with purpose and not at all look lost.

I came to some steps on a columned building, made my way up and entered into an inner area of more columns and rooms. Now I was amongst more Roman citizens, some smiled and nodded their heads while others just stared. I was trying to look as a slave upon a purpose, hoping they were thinking the same.

I continued on until I came to a set of rooms to duck into. Looking around, I pretended to have a task and needing to get somewhere. So I picked up some pieces of parchment laid about and walked right back out, down the steps and onto the same street that I had so recently left.

My supposed purpose seemed genuine enough, but my real one was to just get back to the aqueduct and rethink if I would continue this visit to 320 AD. I was feeling inadequate to maintain my acting abilities without a better command of Latin and behavior for the time and setting. I needed to check on the Hypersphere and felt that an exit may be advisable.

On and on I walked with the parchment pieces that I had no understanding of and yet used as a prop in my act. A few people I had passed previously in the market area were staring at me. I just kept my pace and looked as though I had to return to someone somewhere who sent me on this errand. This had to work for I did not know what to do if I would be detained by a crowd or worse soldiers.

Finally through the gate and back up the road to the linear run of those welcome Italian Cypress trees and the aqueduct. Making my way closer to the base of the structure, but still a short distance to go, I saw an old man approaching. He was mumbling to himself and meandering from one side to the other of this via. He also had parchments; what would he think of me carrying the same? What if he tried to inquire of me? What if he would become irate if I did not understand him and start calling for help?

If he wanted, he could have my parchments for I would gladly give these to him. I had to get to the Hypersphere and back to the twenty-first century, for I was over my head in complications. If for some reason I would have a situation that I could not get out of, then it could mean no return for me from this time and place to my home and time-space, still over 1600 light years away.

The old man was staring at me too, so I just smiled as he passed by. I felt a need to turn back a moment, and found that he had stopped and was looking at me intently. He raised his right hand and I abruptly turned around and kept going in the direction of my shelter. Making my way up the road faster now I relocated the proper base of the aqueduct to re-enter.

Entering at a fast pace, I let my eyes reorient to the darker interior. 'Whew!' I muttered as I could finally sense some relief from today's ordeals. I moved in the dimmer light towards the Hypersphere, but almost tripped over something. I made quick use of my hidden LED light and found that someone had moved in here too. Skins of deer, a clay pot full of maize kernels and bread wrapped in cloths of various types. I wondered if these belonged to the old man I had passed. It was warm here during the day, but the old man might have sought refuge from the chilly nights.

I opened the door, entered the sphere and quickly closed the door. Once inside I could get some comfort and privacy to think about my situation. What if it was the old man? Was he not shocked at a metallic mysterious sphere in his midst? Oh, my! What if this old man is the same old man I had gotten to know in 2010 AD?

It was for me now to venture out, try to find him and discover if he was who I suspected he might be; or just leave immediately and return to the twenty-first century to planet Earth in the future. If the old man was not the one from my 'past present', then he might get alarmed and tell others, to whom it would seem fantasy, especially if they were aware of me leaving this time and space. If I tried to get to know him a little, my lack of Latin might concern him, but that would be nothing compared to opening the Hypersphere and emerging in front of him.

I peered out into the dim lit environs of this stone enclosed storage area through one of the small portholes of the sphere. I saw no movement about. Maybe while hoarding his treasures he could not see very well and did not even notice this Hypergeometric Mechanism before him.

Deciding to kill the internal lighting I slowly opened the door to emerge and wait for him. I was trying to make sure he would not be frightened. Also, this way I could quickly reenter and get away in case of any problems suddenly arising. I had checked the entanglement, and it was still there.

Going toward the doorway of the aqueduct's base I carefully peered up and down the roadway and found the outside was quiet. I ventured again to the row of cypress trees and peered down into the small valley that became part of Rome. I could smell cooking and hear better the noises of the day. Here I just soaked in the surroundings and felt safe so close to the still entangled Hypersphere.

Turning back there was the old man again exiting the dim lit room where the Hypersphere was. He made his way farther along the base and disappeared around it. Now my curiosity about him was aroused more as I was convinced that I already knew with certainty what I was trying to substantiate.

Slowly I made my way back to the sphere and in the dim light approached the open door. Using my small LED light and looking around, I saw a rolled parchment at the base of the Hypersphere just before the door's entrance.

I was positive of the identity of 'the old man from Rome', as I thought of him. Going inside I closed the sphere's door. Turning on the internal lights but not too brightly, I did not want to make a 'display' of the Hypersphere in this dark and safe storage area.

I was exhausted and actually so anxious now that I even rechecked that entanglement was still functioning. "Thank God it was!" my whisper reflected aloud my thoughts within. Inside, I waited, pondering my circumstances nearly forgetting about the rolled parchment that I had brought in and placed near my seat.

With the entanglement I had to initiate the sequence from my 'now and here' and I would be on my way to 'then and there'. For when and where are so intimately related. I was beginning to feel the machine react to the increasing influence of the oscillations and their course within the hyper-geometric chord of entanglement.

If someone came into the dark storage room of the aqueduct, one would suddenly witness the sphere humming and brilliantly squeezing into a disc-like form that was horizontal to the ground. There it would soon be very thin, redden and then fade to just 'gone'. One from this fourth century era would be so overwhelmed that to share such an experience would be considered mad folly. I wondered if the old man from Rome might have watched ... and if he truly was the same old man of the twenty-first century?

In my hypergeometric flight back to 2010 I had over three normal days to read the parchment left for me. Now the consequences of such were apparent beyond doubt; as I had often heard the old scientist say, "The implications of the application of Hypergeometry to practical mechanism are provocative." And so it was. There appeared to be the possibility that with time-space travel, one person may run into another known as once gone or dead, and in another time prior, but certainly kinetically applicable, still alive.

Here was a sample of that probability when the old man must have traveled into the future prior to or at a particular time in the twenty-first century, and somehow was aware of my interest in going to the fourth century. But he would have had to travel to 320 prior to his death. Here was a true macroscopic counterintuitive quagmire of sorts. I must have mentioned my interest in one of our conversations, for we did chat about ancient civilizations including Rome, as well as Egypt.

These world lines between events of the forward only (going to the future) were quite easy to graph; but for forward and reverse transits from past to future, and future to past, the graphs can become complex. Causality protected by the distance of space, now becomes more obvious. With this change of a world line to an alternate world line, the course of the local and the universal events of history continue to propagate forward as a well. The whole event progresses, even though there can be more minor scales of digression temporally.

This is reflected in some life processes like refrigeration being reverse entropic in an entropic universe. As with Hypergeometry's extra energy required to go 370,000 times the speed of light in reverse time, a geometric chord below the easier sub-light speed and forward time above the surface of the greater universe, must exist. At the speed of light, time goes to zero, the surface of the Event Horizon of this greater universe itself floating upon the same surface of the five dimensional superuniverse; the fifth dimension, being Inertial Geometry or rendered by so many as 'gravity'.

In this practical endeavor of machines now able to be predominate in time as well as always in space, one glimpses the grander vista of God's natural worlds.

At one point after napping I did unroll the parchment. I hesitated to even open it, let alone to dare read it. But with solace and zeal I did read it slowly.

In it he mentioned that 'all of this requirement of Hypergeometric Mechanics along with the calculations and resolution of accuracies, the immense use of electric and derived magnetic power is what we Inertially Geometric creatures need. When one dies they no longer need all this. As it is written: "to be absent from the body, is to be present with the Lord" for the believer in Jesus Christ; that is all, as simple as that.'

He later commented that he had attempted to visit the time of the Messiah, but that is when he felt oppressed. When he tried again, he was 'brought low'. Later in this parchment of dissertation he expounded that he had come to the revelation that when one leaves their tent or he says 'vehicle', they are with Him. Then there is no need for such Hypergeometric Mechanism by human hands any longer.

He was at peace with what he had learned. And in his limited human use of the Hypersphere's capabilities, he had wanted to make sure I would get back from the fourth century. So I was positive that the old man from Rome was one and the same as the old gent I had known so well from the early twenty-first century.

I knew it as knowing that I know. He had so profoundly taught me of the great glimpses my amateurish ambitions would allow me. I could agree with Sir Isaac Newton, who said that he had 'seen so far, for he stood upon the shoulders of giants'; shoulders such as I was now standing upon. The seemingly terminated old man was not to be so after all.

Chapter 15

2010 AD and More Profound
Are My Thoughts

I arrived back in my time and place with a bit of the Biblical, proverbial limp of Jacob, for I had wrestled with God in my own way. I was not at all irreverent, but rather naive and bold in my conjectures of so limited experience of life. I had the amazing use of a machine that I had such little understanding of. It was as a child driving a fast automobile, meant for a mature adult.

I admit this time in the Roman jaunt, I certainly bit off more than I could chew. The old man in his greater wisdom had parented me in my escapade enough for me to get back to my primary time-space safely.

I let a few days go by before I began to look through his notes and books again. I could linger here and there, and not get myself into any trouble while sipping some tea or the remaining red wine that we had shared. I slowly took more and more time to reflect what all of this Hypergeometric math and science was saying.

I looked at the plants on the window sill and gave them some water. After sitting back down I wondered if time is of the scales of space, then not only where we are, but when.

He had such a great expository on Inherent Differentiation (IH), when life is a mathematical function which progresses according to only two triggers, time and environment, though each has minors in their set. Here, now, life is a topological geometric function, progressing as a smaller non-entropic in the larger universe of entropy. Here it is a hard/software of a computer designed and maintained by its Programmer-God and it unfolds thus per its inherent possibilities. This seems to imply that the life on or within other planets and their oceans is similar generally, but quite specialized in appearance and function pertaining to its very particular locale.

Carbon-based life is probably the most predominate, not with the disallowance of others though. But if we seem to be a very small sample of His greater work, then on the greater spans of time and space, we may have biological 'cousins'.

The recent work of consensus that says proteins and their selection of only certain amino acids, appears to add substance to the mathematical basis of life. The very recent work of post-selection mathematical functions in logic apparently clarifies what at first seems counterintuitive in the paradoxes of quantum and relativistic physics, and also is revealed to one who bothers to reflect as objectively as humanly possible.

Thus with the unsubstantiated notion of 'randomness' replaced by this more deterministic mathematical approach, we see not only forward and reverse time travel, but also how the greater perspective brings understanding to the limited smaller perception. The quandaries of the Quantum Eraser, the grandfather being killed by the grandson before the grandson was born, and what life is probably like in the many more Pre-Cambrian states of other planets around other stars, are made clearer.

Where in post-selection, with only the true state being allowed, then all other possibilities are false, and so the forward time to reverse time function $(t \rightarrow -t)$ is far more apparent. What this elucidates is as when billiard balls on a billiard table are filmed in mid action, one cannot determine whether the film is running forward or backward. Therefore time is mathematically able to be forward or backward. Life is prolific in billions of years ago and billions of years to come in our local observable universe floating upon this greater superuniverse. Thusly in a limited sense, that which seems incomprehensible in a grander perspective becomes clearly understandable. Thanks to Seth Lloyd and MIT for his post-selection.

God's nature is not limited to humankind's Earth and not for its meager convenience. Rather 'God's higher ways' are gradually being proven so we at least might glimpse them in our limited ways of resolution and mechanism. It is apparent that evolution's non-mathematical foundation with apparent 'randomness' is being replaced by a more scientific, mathematically based determinism. Those who choose to rationalize might by their own consensus keep their 'pet belief' safe, but are left to their unscientific delusion.

On the greater scale of our Galaxy, and deep time and deep space, we see the grander continuum giving context to our much smaller local niches of time and space. Per chance, our limited human historical perspectives allow some of us comfort from testing of our beloved theories. These implications are invalidated in when and where we are.

Now on this greater stage our theater evokes a place not so warranted beforehand. Distance of time and space are bi-directional and thus perspective becomes shockingly clear. More questions are born after others are answered. The more one seems to acquire knowledge, the less is really known. Humanity is humbled, its pride so reined in, that the arrogant seem to gnash their pseudo-intellectual teeth.

Hypergeometry is a perspective that truly upsets the paradigms of temporary human establishments such as science, mathematics and even biology. I was so humbled, but fascinated by it all. I had to put the books down, finish my tea or wine, then leave his apartment. After such a period of study I needed to rest and digest quietly without additional input. Then as a distraction, I had to immerse myself in daily living, enjoying pleasant and simpler entertainments.

I listened to some old songs, a few just instrumentals, and got lost for a time in them. The quiet was wonderful, and I was slowly going the way of things on this Earth, getting older. I thought I had quite a bit of time yet, for I was much younger than the old man. My prime time of 2010 AD, which is mentioned colloquially, was upon the Lorent-Fitzgerald Contraction for Time Dilation offered as time-origin, where the dilated time of the result was labeled time-prime. This I hoped would bring some clarification to the old man's and my referencing of such.

In his study one day, I discovered in his papers that he had actually been to the Coliseum and Circus Maximus in ancient Rome during the fourth century period. This surely brings closure to his mysterious appearance when I was there.

How complex is the inter-entwinement of world lines of the temporal consequences when they are the predominates in traverses beyond the 3D of space and one dimension for time. And to consider the presumption of another, a second dimension of time. Some physicists have offered that increases of complexity of another ninety degrees, temporal transformation does exist for space at the spatial 3D. If time has a second dimension, it may also have a third. Could such occur by manipulation of inversion from our 3D space and one dimensional time, to 3D time and one dimensional space? But consider if such is contemplated, we may have already through such only just arrived at the end with the other side of grand superuniverse's Event Horizon. Our Tachyon World mantle to us as occupants is to them who may occupy the other side, reversed and so results in now another 3D space and one dimension negative time.

Such reflections, thank God, we are so able to do. But I had to take a sip of the wine the old man left behind, for I could easily get heady in all this. I was ecstatic to simply read about and appreciate his own journey to ancient Rome, in some prior time, maybe a year or so before my 2010 exploration. What is intriguing is that he had left some time before, yet we had met in 320 AD. How complex the temporal implications had become.

It appears he was at Circus Maximus and the Coliseum as I had supposed from my earlier perusal. He surely made better progress than I and I felt humbled at his aptitude for Latin too. I sensed that he struggled with the locals' dialect, yet his notes were able to hold me fast in interest.

I would close my eyes after reading some of his notes, for at times they formed into such prose that glimpses of epic proportions enveloped my mental theater. Here he was now writing as a novel, but with such details as a scientist-explorer, that I could not escape feeling as though I was right there with him in his struggles.

He reminisced of one particular incident when he had made it inside of the Coliseum as a local. He somehow had a drachma to buy some bread being sold amongst the animated crowd in an almost full arena. Here he was ripping apart his bread, trying to maintain his seat and his disguise as the tournaments were being contested before him upon the great floor.

In the battles of gladiators, for instance, the raucous screams and chaos of the fans were deafening. Fights broke out in the seating areas and some were injured as well. Roman guards were here and there, and some tried to maintain order, but for the most part it seemed that if it was vicious entertainment one wanted, then vicious entertainment one may get, even personally.

He writes of another moment where he was in Circus Maximus. How did he get around so? I paled in my own lack of ability in Latin, never having realized the need to know accents and proper dialects for the time and place of that ancient Rome. There he was able to witness a chariot race. I became more amazed at the adaptations required. He may have been within an inch or two of his life at some moments.

This particular oval arena of simulated conflict and sport allowed the re-creation of Rome's wars displaying their victories. Each of the amphitheaters could even be flooded to portray water-filled 'seas' and 'harbors' to show such military prowess of the empire.

It was of more fantastic ingenuities that from amphitheaters to areas of theater or sport where the ancient technologies required were quite adequate. In these two arenas of ancient Rome, we are sure that lifts were used to raise and lower the central grounds allowing the entry of an army unit, gladiators or wild animals, and sadly even prisoners to be executed, often en massé.

I had to admire the old man's daring for he surely had studied and equipped himself well. His ability to mix and make inroads within fourth century Rome was a grand achievement. It made me realize how short I had fallen. I also could better understand that with any temporal endeavor embarking in a deeper time episode, I would have to more proportionally manage my own preparations.

Just consider the use of the English language. A journey of even one hundred fifteen years in the past with the same in distance, words and phrases would adapt and diversify albeit slightly enough. Such as 'fancy' in the latter nineteenth century meant 'sporting'. Going farther to back the Elizabethan Era, our modern English language would be at times difficult to get across and for them to understand. Farther back in time, then we surely would be more the proverbial 'stranger in a strange land', as was bespoke of Moses in the Old Testament book of Exodus.

With distance, both temporal and spatial, the traveler would be prone to the unfamiliar and riskier of environments to blend in to socially and safely. Given some similarities in faraway shores and cultures, one usually finds the less known and then accordingly more to learn. The same with far away times, whether past or future, weeks or years, one finds adaptations demanding much more than the near. Let's reflect a bit upon the future. Now dare I to meditate upon one thousand or one million years. It is almost as if we had gone to another planet. Such sporting is surely daring and even dangerous. Now the same back such years, what consternation and wonder in ambiguous relation would one certainly experience. I had an enjoyable and adventurous time in the nineteenth century. But in the fourth century, I began to stumble in my efficiency of adaption so much, that I knew I had bit off more than I could chew.

Looking at geometries of the simple shapes of circles and triangles inspires more considerations of the time and space I seem imbedded within. A triangle has one hundred eighty degrees in flat space, if convex then the angles within sum to over one hundred eighty degrees, and if upon a concave surface then to less than one hundred eighty degrees.

This surface I find myself upon, with light at 370,000 times slower than the rate of time in the accompanying geometric chord 'beneath' it in the mantle of the Tachyon World, then the chord is always 1/370,000 the time it takes, but in reverse time.

How such simple expressions of plane figures seem to reflect a projected higher geometry. Most circumferences are not of such a greater magnitude to a chord. But in this case the chord for this part of the circumference is *2.7027027... x 10 e -6*. This may imply a higher curvature upon the equator of our greater mantle where our known observable universe is floating.

Again, locally all of our circles and triangles appear within intuitive restraints, but enlarge them for light years and thus increased resolution would allow such earlier reference of greater than one hundred eighty degrees of summation of the triangles mentioned and the squishing far more of our circle.

Such indications are so slight in all forms of practicality for some distances of time and space, that it is assumed unmeasurable, and non-observable, and so abstract and esoteric to be fanciful.

Such choice of ignorance thus forfeits a grander and so much subtler revelation to the human senses. Einstien's Block Time that so appalls many, when the past and future are as real as the present, now can better with sincere interest be evident to accept. What of the 'gravity' and the inertia of different parts of the local universe laying on slightly differing curvatures upon this super-surface of the superuniverse?

I looked again at the figure from the inversion 1/370,000. Its equality was *2.7027027... x 10 e -6*, which made me think. Is this not close to the Curie (Ci) of decay found in radiation of matter? Is this relating to our observation by measurements of a radiation background? Hmm, intrigues propagating intrigues.

Upon the circumference I again contemplate that to maintain it spherical, am I measuring the greater part of the total circumference to allow for the ratio I had mentioned just beforehand? If it is the diameter, then it may be considered a constant. It would make the superuniverse more spherical and perhaps better to analogue as to understand the geometric projections I offer myself, and anyone daring to overhear me.

All this heady indulgence had me seek rest and sleep. For the practical man that to some extent I am, then as long as it worked well enough and was safe, I would not mind going for such rides. My favorite so far was surely 1895. There I could have stayed.

The old man had related of how his father laid disheveled on his bed in the night as the after image light passed through his bedroom in one of the early tests of the Hypersphere. Though he was deteriorating in his late eighties, his father had new insights into Hypergeometry.

After being tested by his personal doctor for dementia, when he could not even draw a sketch of a clock showing four o'clock, he was put on pharmaceuticals to slow his loss of normal cognizance. But at this same time his father had begun to show savant abilities in the physics of entanglement. The doctors dismissed him, but his mind seemed to have far more powerful perspectives. He, along with his son (the old man) had theorized and developed a machine that would be able to obtain 370,000 times the speed of light, whilst according to the norms of society and the medical field, he was considered useless.

While his doctor professionally categorized him as demented, history would now record that he and his son would achieve mechanism far beyond the understanding of his critics. It would even challenge the physicists of the time and in years to come.

Such abstraction when actually mechanized reduces the theoretical to practical experience. This was what had astounded those who had somehow experienced such in actuality or witnessed the repercussions of effects in their environment of these experimentations. Lost in thought and pondering is what I seemed oft to do with more succession, whether of actual practice in the Hypergeometric of mechanism or in contemplations of such.

It was easy to try to maintain the routines and practical everyday life of my primary present, but then to drift off in thought. Even absent-mindedness for the moments that gentle winds pull or push to the side that which had to be paid, or repaired and maintained in domicile or automobile. I soon discovered the containment and calmness like the old gentleman. It is written that 'a rich man is content in what he has', and how true. Somehow this endeavor of temporal transit was also a comfort.

Somehow to be local in one's life geographically, yet also travel in time predominately along with space, is of such wonderful interaction. The fantastic ability to journey also great distances of space and time so much that the counterintuitive of remaining locally entangled generated a quietness. I no longer felt any hurry, no more race and chase. I would not say that there were moments for anxiety, sadness or disappointment, but rather there was a closeness of God that made me realize a better perspective. I would try to rest more and I have so well learned that to travel too far in time or space requires more proportional preparation for the needed assimilation. One needs discretion, the better part of valor, as it is often said. The greater the abilities one is allowed, then the greater responsibilities also are of importance.

I find myself more tired with an ever so slight progression of age, needing naps like my old friend. Perhaps I am becoming more like my mentor, whom I admired so much. When left to its natural condition of the normal space-time curvature, time continues to course from the past to the future at its local rate of one second per second. Simultanity seems at these nominal bends to allow one to easily think as humankind has often thought of time, as merely interesting abstraction. 'Now' is sufficient for all practical purposes.

Again I contemplated the two small discs that were at the old man's rural laboratory. They particularly intrigued me for some reason. I had intended to go back, pick them up and study more about their significance. He always related that they were a subset of the third of the three machines of Hypergeometric Mechanics. What he meant was that they somehow go along in their function with the Hypersphere itself. He had often mentioned that the common camera and the tape recorder as well were 'quantum machines'. I wanted to put these seemingly minor subjects under more enlightened scrutiny. They did have some purpose, possibly all in conjunction with each other, I'm sure.

First the two discs were resonant to the actions of the Hypersphere for they were inertially 'in tune' with its activities. They would oscillate in harmony with a passive response. The sphere was the active driving element while the discs were both passive to it. Using a camera or an audio recorder with close proximity to either disc produced an interesting local condition for the recording, Non-intrusively of other times.

The discs could be used by the holder in their pocket to affect by inertial sympathy their very local environment. This could well be part of the elderly gentleman's experiments that he had noted in some of his journals and had related to me, for he claimed that there were ways to view and hear of other times without inertially interfering. There was some energy leakage from such uses, but amplification was the way to go in order to allow observation and measurement of those very weak, advanced waves that James Clerk Maxwell's Equations gave equal validity to in symmetry of function.

But the surprise for me was not only the electromagnetic response of the photons, but the acoustical-mechanical phonons as well. For it seemed with the discs that a past or future time became a subtle replication as an experiential environment for the disc user. One could photograph and record sound though requiring actual undistracted focus of human and mechanism.

This referred me back to the use of the Temporal Diffraction Grating, the first of his three Third Level Machines. Not only measuring the direction time initially, but its use of the appropriated 'now' or 'present', when inertial/kinetic measurement was done, to image the slight 'befores' and 'afters' in the scale of partial seconds of any photograph placed upon it.

Thus the two dimensional camera and the one dimensional sound recorder were equated to 'quantum machines'. Here one can observe while not kinetically interfering with that temporal and spatial environment that otherwise is called world lines. It is as if one has a portable window to carry and able to peer through to a time and place long gone or yet to come. But the ones in the future are very brief or far more difficult to discern. First of all, from future to past is the weaker signal, such that the signal to noise level is very low, then the reverse. That is why, perhaps it is the past that is old enough to be remembered and of effect. The signal from the past seems stronger, and we of the future have so minimal a kinetic interference. So then the same for the future signal, all this per Maxwell's Equations, for the retarded waves from past to future are similar in respect to the advanced waves of future to past. It is as we see the past in the night sky, but cannot see the future from the opposite temporal direction, though it must be there, ever so diminished in signal.

It is also interesting to note that if one views along the time symmetry formula $t \rightarrow -t$, then all we have to do is extend the line across a piece of paper horizontally. We can map as a diagram the flow of time from $+t$ to $+t$, or $+t \longrightarrow \longrightarrow +t$.

Next make a small dropping perpendicular sinusoidal line emanating from a point upon this line also going forward to $+t$ and in parallel, then perpendicularly bring this wavy line back up to the one above it going forward; you have a schematic of forward time travel, with the wiggly line representing the extreme Lorentzian effects involved. These effects as foreshortening in the axial line of oscillation, increase of mass, and the dilation of time on this shortcut to the future. The line then returns to the present time of the stoppage of oscillation and you are back upon the horizontal straight line but in the future!

Then take this same single line and below it add another line but label it times two or whatever would help to designate it appropriately, and have it going in reverse so that it would be $-t \longleftarrow \longleftarrow -t$. This is how it would look in reference to our flow of time. All here is relative of one to the other; but again all is relative in this reverse time if one is imbedded within it. It all seems forward, and we are in reverse.

Now draw another perpendicular line to the one from the forward line above and using another small sinusoidal but broken line (to indicate Non-inertial effect of passing through an Event Horizon), actually just a smaller part of the superuniverse's Great Event Horizon, which equals the speed of light (*c*). This line drops down to the reverse line (the times two line) and is riding its new inertial state in a reverse time direction to us; but to itself and all there, in their forward time, to our farther past.

Lastly draw the Non-inertial broken squiggly line perpendicularly down to the times two line (reverse time), allowing the past to be accessed. This wavy broken line should be angled to the left some, to indicate that one has also traveled at 370,000 times the speed of light (*c*). But it is understood, for the line on the bottom can be adjusted to the right accordingly to represent an appropriate past and distance in space.

To make the concept easier, let us just focus upon the temporal. When having arrived in the reverse and going in a reversed direction (extreme Lorentzian state) in relation to our normal time of *c* times 370,000 in the Tachyon Mantle of the superuniverse, we then surface from below the Universal or Grand Event Horizon at some local point; now in our own times past. This is represented by another perpendicular broken wavy line until its juncture upon our own timeline's past. Then we must continue above our normal time line, for we are suddenly in an alternate history; so any effects from us will not affect our home time line's history.

This all so heady, but graphing it allows us to better see it conceptually. The old scientist surely had lots of these time diagrams schematically representative of how alternate histories are protected causally from the other in the superuniverse.

He would say that 'humanity was in the 'Hypergeometric Age' while so few realized it'. So then I wondered if maybe we had been in a more wonderful time, perhaps in Eden or the Pre-Edenic Era. I surely could drift off upon such for some moments. Why even the ancient Greeks were fascinated by those of more ancient times to theirs. Such fascinations of temporal and spatial, history and myth. Now it was time for that wine, for I was worn out after such thoughts. In a little while I had better pursue those two discs he had left behind to see what would avail for me of their use.

I had slept in his old chair, again and I perceived myself following his path more and more, even greying and becoming more content. I soon remembered to pick up the appropriate journal about those discs and placed one in each of my coat pockets. I would study them and see what their potentials were for my own curiosity. It is so fascinating to contemplate much grander things, it even begs the question if our understanding of reality may have holographic indications. It may be more like that in the temporal Non-kinetic interfering, for then one is 'virtually' there while yet unable to make any effect on that observed environment. Consequently it does not affect one's home world line either and is a far more complex method of looking at old photographs and listening to old recordings. But the effect is more like a 4-D shadow, but very amplified.

I planned to try an experiment in the near future, but I had yet to work out the details. Remember, I was still learning. I would initiate the Hypersphere, but just to oscillate and become entangled to it by one of the discs. I had the idea that it may help me to more clearly understand their possibilities with the Hypergeometrics involved. I did not plan to travel anywhere or to any different time, but just allow the sphere to run in oscillation as I would carry a disc in my pocket. I knew from his writings that if the sphere was modulated in patterns of oscillations, the changing amplitudes and frequencies would affect the two small spherical weights inside the discs. But there may be other possibilities as well.

I had also read that the local space-time around the discs would be under harmonic curvature effects in resonance to the Hypersphere's oscillations and their changing patterns. Seeing as both are in resonance, each was in phase, and became entangled to each other. So my local physical space-time as well would be in phase and entangled. I wondered what other articulations could be done. There was a complex portion of his work that seemed to say that one could pass through a wall, not go through, but yes, 'pass through'. At some resonance, there seemed to be an odd effect such as this. As a matter of fact, he mentioned one of his granddaughters had gone through a mirror years ago. Fascinating! Now I wondered how long he had been doing such things and how far had he really gone?

I would have to make the proper preparations for the equipment involved. It would be some days, but a moderate attempt would be quite the icing on the 'cake' that I had been ingesting recently.

In his journal, I had noticed a curiosity ... an old piece of paper on which he had written: 'Mom, maybe I shall put a hole in the sky'. He continued that it was something that he would try to achieve. He had told me once that his father had asked him if he could 'build a machine with one moving part?' He tried to accomplish this task as well. It seemed that all of this had motivated him, because he also noted, 'as long as no one has proven that I can't, then I can'. This was how he thought and planned in his mechanizing geometry. The Hypersphere certainly had one moving part, and the hole in the sky must be in the 'sky' of time and space. Interesting contemplations I surely indulged myself in.

I took the discs with me for they were not of complexity, astonishingly simple. So simple that if one were to just find one or both and look them over, they would be considered some odd type of toy. They were not powered by any electrical source, and they were essentially sympathetically resonant inertially to certain frequencies generated by the Hypersphere. All they really did was oscillate in harmony with the main frequency of the sphere as it shook the local space-time curvature. This tidal-like effect was affecting the local area of space-time and anything resonant would energize inertially from the input of the Hyperpshere's inertial harmonic oscillations for miles.

The Inverse-cube Law seemed to be best to describe all these kinetically-radiated, emanating waves. Consequently if these discs were both precise to certain resonances, then they would effectively not only oscillate, but also rectify any modulation coming upon them.

Now, as I understood it, the local space-time could locally be of interesting consequence, especially with a camera or audio recorder. This would also affect the person and objects nearby. It would be as though a small bubble of predominate time-space anomalies were occurring. One side of each disc was transparent, maybe quartz, while along its rim was like polished brass. Its opposite side was opaque of the same brass material, but inside it was hollow. There may have been more inside the disc than I could ascertain, besides the two small metal-like dense grey spheres connected to each other by a spring. I must say that I am guessing a bit at the materials and all their functioning together, but it was like an Inertial antenna to a great extent. What else these small contraptions could be capable of, I did not know. But I ventured enough to bring a camera and initiated the Hypersphere. But this time I would not be nearby, but quite some miles away, the distance of enough responsive affectation he noted was around three hundred miles.

These discs seemed easier to understand than the Hypersphere itself. But they were adjusted to the sphere, for it was their driver as the oscillating magnetic field in opposition to the constant local 'gravitational field' or as he would say, 'the local Inertial Geometric curvature'.

I needed to study very carefully his notes, but there were so many scribbles and corrections as to make my task very formidable. Was I to carry both of them or just one? As I scrutinized his theories from notes written on napkins, backs of envelopes and well-organized typed journals, I was slowly gathering more and more understanding. Some were so difficult to read that I had to use a flashlight from behind, always adjusting the light to make out finer details otherwise just barely discernible. He had always seemed so hurried, as if the idea was to be captured precisely at that moment of conception. I pity anyone in conversation with him at such a moment, for surely he would have left them behind as he escaped to his latest idea. I can understand better how some of his family or friends would find themselves adrift and alone from his social interest.

I surely was hoping that in my next journey I would have enough knowledge to be sure of what I was doing. So that instead of some egg-on-my-face, or worse an injury or death, I could actually achieve something of great interest, and really 'enjoy the ride', as the old man would often say. If in this journey, time was predominate and space less so, then I would rather to make sure that I did not somehow get myself too far back or forward.

I cared not as much to journey into the future as to those years of the latter nineteenth and early twentieth centuries. They were at that time looking to their future and its possibilities of wondrous advancements in science, medicine and technology, and the fulfillment of all their imaginings. I was from a time when such had actually come to fruition bringing more revelations of extrapolations of engineering, but also regrets of how the economically and politically elite would render such for their own gain and the destruction of morality and freedom. As they longed for the future, I longed for the past.

I would have to give myself enough time to better understand what I was doing, the details, no matter how tedious, confidently within my grasp. I did have a specific time in mind that would not be too far back. There were trollies in the early twentieth century and it would be interesting to maybe go to 1905 or to 1935, so I could notice some progression. I had gone to the time of my great-

great grandparents, so perhaps I could visit the time of my grandparents and parents. I did not want to interfere, but just visit. If that did not work as completely as I was hoping, maybe just some images or sounds from that time would be the most Non-intrusive that I could do kinetically.

The danger in visiting was to have already so slightly interfered with history. I wanted to visit and yet return to my home time. I also knew that there are past times so much more comfortable than my present one. But then every past has its injustices and tragedies. Especially, seeing those so disabled or diseased, knowing that one cannot dare to intervene. It would even be worse to see a family member suffering somehow and not be able to interfere.

I had read that the disc camera was able to take pictures and even record sound from the past. I could perhaps just take some pictures with my camera from modern time or merely be able to see and hear some things while not upsetting the course of history.

There used to be trolley lines connecting small towns years ago. Tribes Hill, NY had a coal-fired, electrical power station and there was a trolley line from there into Amsterdam, NY. What a safe trip that could be to just anonymously take a trolley ride in the 1930s, right now just a thought, but so intriguing. I should be able to get something to eat and watch the people for a day.

I remember that the old man had said so often, 'It is written, that we have a cloud of witnesses'; so we must always be aware that in time all things are seen, heard and known by others, and 'already beforehand, during and after by God!' This used to bother me for I could not handle injustice, or what I perceived as such, even when later it might be due reward for someone that had perpetrated an injustice. Why a baby dies (when it is God's first and the parents second) or an innocent person is harmed for no reason. Time makes the understanding easier when there is a bigger story not yet revealed. My personal faith in God has increased with this temporal perspective.

Justice is limited by us humans for we only see the myopic here and now. So we are able to do what we should do ... something to protect or help someone. But when one is in another time, it is as if on another planet. If this 'grand program' is followed, then we are alluding to a 'Great Programmer'. Same with time, for it is not moral or ethical, even amongst atheists, to interfere with history's progression. Why? From a Petri dish of bacteria, to another Earth-like world in another solar system, to another time in our past, we all seem to accept not to interfere.

No scientist would ever want to affect their own results just so they could know what the consequences of the variables would be. If one interferes, then their observation and time learning under certain conditions would be for naught. For example if a scientist wants to know if a rose seed will thrive in the Martian environment, then the experiment must have an environment as similar to Mars on Earth as possible, or have a spacecraft actually land on Mars, plant and then observe the seed in actual Martian regolith under actual Martian conditions. If one interferes to save the rose, then no one will know if the rose seed would actually thrive there.

So interference hinders our ability to learn about incidents and their consequences within human history. Should one prevent the assassination of President Lincoln at Ford's Theater? Or would it be better to learn what actually happened and not to interfere?

It appears true that the distances of space and the distances in time are proportional. The farther away in time, then also the farther away in space, and so Hawking's Causality Protection Conjecture is well assured. But this also alludes to God's greater protections, which are not conjecture that we may only glimpse or not understand at all. If we are only able to discern such things, while they have always been, then we have to admit that things we do not understand still work well, whether we agree with Him or not.

Humanity only discovers those things that are already there. We saw the backside of the Moon in the 1950s for the first time, while it was always there. By 2010 we had learned more about Lincoln's assassination than the general public knew at the time it happened. People in those times were aware of some details they thought were undisclosed, but we have discovered some of those secrets that were always there. If Lincoln was not to die that night, then God would have prevented it and would have programmed history accordingly. If one does not want to know how a cow is made into a steak, just eat and enjoy your steak. But if you want to understand the process from cow to steak, stay away from a cattle ranch, the Chicago Stockyards and the butcher shop. Just go into the insulated world of a grocery store and buy it in oblivion. There are those who can handle some things, and some people who can manage different things, and others, not at all. But that does not mean that they can handle the same things. As with beauty, one's justice is another's injustice. We all see only our own side, and no matter how we enlarge our perspective, we still are limited with justice on the greater scales of time and space. We are visiting in God's area, and we are to be interfering only in our own designated time and place of the present here and now we are living.

Time and space travel has humbled humanity to its proper place. This impedes those with egos who desire to be their own 'god' and conflicts with the 'gods' of others so that they feel the need to oppress. They have the same problem that Satan in the Bible had, an arrogant pride that defies God. Time travel and space travel become religious experiences the farther in distance that we go.

Perhaps the early twentieth century and Tribes Hill, NY, I had finally decided after much pondering.

Chapter 16

What the Two Discs Can Do

I had attached a small digital camera on a tripod for stability with the Hypersphere oscillating as the old gent had said about 300 miles away. I had investigated and found that the discs could be used in unison forming an Inertial/Kinetic interference receiver that was bi-directional. Just one could be used to more omni-directionally affect my local environment. It was this latter configuration that I would experiment with.

I remembered that he had offered that 'space is recorded in time and time is recorded in space', interactive because both are never still at all known scales of resolution. All this based upon a Hypergeometric mantle underneath that supported both interrelating functions. This included alternate histories as well. Each world line required some substrate of form to exist upon. According to Maxwell's equations, both advanced and retarded waves were valid; it was their relative perceived strengths that differed so immensely. To overcome the far more pronounced retarded waves of forward duration, which we experience normally, filters and shielding were required to minimize this and also cryogenics. On the other hand to enhance such very weak advanced waves required cryogenics and some form of very specific amplification. The resonance of both of the discs or a single disc to the distant Hypersphere, acted as an amplifier and filter together. Enclosed in this special system were other recording equipment which one could with some more unusual engineering, Non-inclusively record past or future, while minimizing the present;

remembering what Einstein mentioned in his Block Time concepts, that past and future were as real as the present.

The kinetic signal to noise ratio was highest for the present, such as we experience profoundly being able to kinetically interface with our environment. The past is a somewhat weak signal and the future is far weaker, but in reverse duration. This is the basics of retarded and advanced waves respectively. Therefore a signal from the past from London to New York would be much more readily receivable from past to future, for the signal may be only milliseconds old, considering the speed of light in our forward going time. In contrast a signal from New York to London, in reverse time, thus an advanced wave, would be far less detectable and be received milliseconds before it was sent. This is far more profound with greater distances in space. So a signal from Alpha Centauri is 4.32 years delayed at a certain strength, while its signal going backwards in time would have occurred much more weakly 4.32 years before it was sent. This presumes the distance of 4.32 light years between Earth and the stars of Alpha Centauri approximately, for this discussion.

So for my small local experiment in Tribes Hill, NY near an abandoned trolley line south of Route 5 near the local cemetery, I set up in some seclusion a very quiet but profound contraption for this proper mixture of cryogenics, amplification and filtering. Within an insulated black box I placed the digital camera and one of the discs. I aimed at the tracks and let the shutter open and close several times. It actually was just an advanced form of the old man's Temporal Diffraction Grating Experiment, otherwise more colloquially known as, 'The Where is Now Experiment'. Finding 'Now' also established 'Past' and 'Future'. For any photon source is from the past, while its anti-source so to speak is from the future in reverse time. To look at the night sky means looking into the past, so you now have a sense of the arrow of time cosmically. Locally it is just more concise. At this point I had to adjust remotely the oscillations of the Hypersphere and the resonance of the disc.

I had also recently learned more of this in my further studies, with my occasional glass of red wine to sooth the counterintuitive stresses involved. Often it was so much better to just accept a premise and run with it, than to verify with results that better clarified what happened. There were ways of placing the disc and turning it so as to adjust its response to the Hypersphere. It was amplifying by harmonically increasing its internal energy because it was passive to the Hypersphere, which was the active part of the system.

A few days earlier, I had looked at what was in the camera's memory chip. Some images were blurred, while others showed clearly tracks and a trolley going by. Others seemed a bit older, from the early 1900s I had deduced. Then there were some from a later time that clues in the images of the progressed differentials suggested the mid 1930s.

The old man had photographs and even recordings from 1895 and other times in the latter nineteenth century, which he had kept in his private archives. For sound, the same had to be done with an anechoic chamber around the microphone in this box, but in an acoustic-mechanic sense. He had boxes of different sizes and used only liquid nitrogen, which was inexpensive for all these types of Non-intrusive 'visitings' of past or future. As far as I knew, his interest in the future was never satisfied, lacking any evidence to such.

When I finished my interesting work, I packed up everything and headed for home. It was getting cooler on this November afternoon, and I was so eager to see how things had turned out.

As soon as I got home I completed my experiment. I must relate the personal effects of what I had suddenly realized and qualified. When I allowed the disc to fully appreciate its harmonic resonance in response to the Hypersphere, it floated gently, almost unhindered from my hand. I felt overwhelmed as it hovered there in the space above my palm. I found myself raising both hands above my head and praising God. I had experienced what as far as I knew no other human, save for the old man, had ever experienced ... a freedom and peace in time that must be reserved for those like us who dare, just by faith to undertake those things of God so beyond the reserved reason of common men.

I had been bullied, as many who are different of thought and reflection are. I had been rejected by my family, as in all these things my old friend was. Now we two had something set aside for those of us receptive and submissive to Him Who has made all things. It is a religious experience to be found in the Hand of our Awesome God.

I realized that we had crossed into what others of humanity would categorize as the Hypergeometric Age, but had already been predicted. It was an epiphany and wonder at how such a despicable, but willful person could rise into new times and worlds, but by the Hand of God who was the Creator of such higher things. He had abruptly dropped me to the ground as the disc hovered over me. I did not even have to be in the Hypersphere, for this was a local anomaly around me for a small distance.

Suddenly I found myself in 1905, smelling the air and hearing the sounds as I lay upon the ground on a summer's day in the early twentieth century!

All of a sudden I fully recognized the potential and capabilities that the old man had achieved, and fortuitously had left for me to also benefit from. It all hit me at once. I was so astounded that I just fell to my knees, humbled before God. For there was no one who would listen and no one who would believe me. I might as well consider all that this had accomplished as a secret to keep for future generations. To think that this treasure was given to such as the old man and me ... we who were so forlorn and unlike others, trying to assimilate, yet in our profound expectations and persistence, we stumbled upon this.

How do I describe finding myself laying upon the ground in the past, or for that matter any of the other of my exploits or of the old man? How do I relate to any other human being what has been done? I have seen the past and it is so pleasurable and familiar, but I do not care for the future, to know it. I can be a silly romantic for such things and not care at all of those who cannot understand. They cannot go, where they believe they cannot. I, as the old gent, was willing to go in time and space where no one has proven that we cannot.

What had happened I reasoned as I sobered some from this sudden ecstasy, was that resonance had been readjusted from 1905 to 1935. Next I must have disappeared from 2010 to appear in 1905, then vanished from thence to finally reappear in 1935, all quite quickly enough so as not to interfere or upset anyone in those times. I had to slowly allow myself to grasp all that had transpired.

It was now as a whirlwind of an experience, unable to be fathomed for quite the while. I had no idea what the consequence would be back in my home time and could only feel comfort holding the disc in my hand as I slowly got myself up off the ground near the tracks. The ground seemed to slightly transfer the sound vibrations to my feet and legs as I was hearing the approach in my ears.

To my left was the Tribes Hill Upper Station, a small building, more a ticket office with some seating. On my right was the local cemetery. All around me were the early twentieth century's sights, smells and sounds. And I was breathing in this time so alive, but in the mid twenty-first century, gone from the earth.

What the old man had done was so fantastic of possibilities, that I could not express coherently any kind of refined comment.

Presently I was just casually, yet incredibly taking it all in. I even recollected that once he told me he had destroyed the first writings and drawings of Hypergeometric Mechanics, and now I understood his concerns. Could such wondrous capabilities be trusted to the greater percentage of humanity? Governments were certainly so corrupted by power and with their immoral and unethical stances throughout history they dare not to be allowed such technology's discovery or usage.

Perhaps after all, this was such a treasured gift from 'above' that only a particular few were considered responsible enough to exercise and safeguard such wisdom. This was a very easy way for experiencing temporal advantage without the isolation of the Hypersphere, but it was because it allowed another derivative to follow the world line of an event. What was amazing was that I was not really kinetically interactive, yet still protective of the causality along this very same world line, almost like in a superposition state. For some unknown reason I was able to receive and sense those environments enough and even stand up. But now I was also standing upon the exact same ground and was actually more ethereal in 1905. So my Inertial/Kinetic interference was just enough near the noise ratio as to be just barely of any effect upon me or my instrumentation. It may be that because the brain is like a quantum/relativistic machine, it may be the elusive factor to additionally consider. The old man had spoken of such theories, and I can only continue the conjectures. Beyond that, it was again, 'quite a ride'.

I had some time after my return home in which to look over his notes, especially in his book, *Hypersphere ... A Journey at the Speed of Geometry*. It encouraged me to take another look at his early measurements of past, present and future, this time in better comparison.

The present is the far stronger in its Inertial/Kinetic ability to inter-react with things over the cosmic background noise ratio. The past was next in strength, then the future, which was the weakest. In his notes he initially ascertained that the past seemed to register stronger. But now I could understand that his instrumentation might have been a factor here. For the time of measurement was a bit biased in the forward duration of present and past, while the future was negative in duration. So though the most difficult to measure, it seemed to be stronger than the past at the forward time of measurement, for it was 'approaching' while the past was 'receding'. In further experiments he did find that the future, when nullifying the internal bias within his instruments and even his nervous system and brain, was the far weaker. We seem to experience psychologically, at about one and one half seconds or so, and because of the persistence of our retinas, auditory nerves and brain functions, we somehow perceive the future 'ever arriving', with the past 'ever leaving'. The present 'Now' is when we seem to be 'staying'. So similar to a Doppler Shift experience, until quantified with all biases removed, the future can appear stronger, but all along it is the weakest of the three signals.

Since his revised edition of the original Hypersphere 'manual', his experiments continued despite his growing more feeble. He had remained active until his passing away. These newer revelations would later come out in 2012, after more of his experimentations were corroborated.

Maxwell's assertions were valid when quantified as purely as possible. The strongest signal is the present, the next in lesser strength is the past as it appears to be going away and so the weakest of all is the future, though it appears to be coming upon one's senses. Even the function of experience is revealed more with the consideration of time as a predominate interest over space.

I returned to normalcy of life by going home, taking all the equipment with me. I was beginning to find myself more like the old man, in contemplation of all these things. More refinement was the result after my experience in all of this for me. I chose to take quite a span of time to just rest. After about a month, I returned to the old scientist's residence. His granddaughters had kept it maintained, before deciding to remove his things and prepare it for sale.

One day one of the girls stopped by to let me know that in another month all of their grandfather's personal items, including his scientific journals and mechanisms would be officially and legally transferred to me, if I would want them. Of course, I would. She and I shook hands and I signed some papers for his things. In the next week I would have to move everything to the rural farm laboratory outside of town. He had also willed me these things and the property where the Hypersphere and other larger machines were kept.

After the reading of the will, all things were in their assigned proper order and place, and already I was back to studying his writings and inventions. He had donated over ten percent of his assets to the Church and the rest to his granddaughters.

By the summer, I had a gentleman's farm, really more a laboratory, and a house to live in. This humble place, safe from outsiders' view, held within a most powerful set of machines that as far as I knew had not been invented yet. At least, the civilization I was living in had not bothered to pursue a quest for such.

I had heard that early on President Nixon was one of the larger influences in hindering human spaceflight. And this is after the United States had taken six pairs of astronauts to the Moon and back safely. Human spaceflight, along with robotics, could have done far more to prosper our economy and expand the human experience. Then we stopped. Apparently the powers behind the throne were now dictating our new capabilities to what they considered more self-serving means. We lost our lead in space and progress in order to build a better 'war machine', tyrannize our citizens in a neo-feudal oppression of control and make for themselves more money. I call them the 'empire builders'. They buy justice, politicians and even governments who then steal the new fruition and kill its possibilities.

The dumbing down of the people by media and politics, so that all these neo-serfs could do was buy things and consume without restraint and be led to believe they owned something. When it came to 'eminent domain' they actually owned nothing. The control of votes and the illusion of the electorate to any form of representation and power were also well contained. The constitution, the authentic, original foundation of the United States, was now in the way of the 'unconstitutional' hierarchy of government within the 'United' States. At this point, even the People realized the Constitution was just a piece of paper in the way of the hidden elites' objectives. Question anything and be labeled a 'terrorist'. I did not want this kind of life anymore. This was a time these arrogant, empire builders could not control. Their drones and surveillance were limited to near Earth space and sky.

America was gone, and the few of us left to realize it, had to have other plans. I was concerned that it would only be a matter of time when they would come to my little farm to steal or even destroy all that had been developed. If they could not handle the potentials of space, and tried to abuse it, then what would they do with Hypergeometry?

I began to wonder as Fermi's Paradox asks, 'Where are they?' ... aliens. Then I could well ask, 'Where are they?' ... other time travelers. In God's great wisdom, He had foreseen all of this to come. It is no matter when in common human history, we are in charge, that He has only a very few able and willing to do these greater things. As in space and in time, He does not reign in second place. Humanity has only this Earth to dominate, and even that for just a certain time.

One afternoon, as I was sitting on my porch, and having not used any of the Third Level machines' technology available to me, a drone and two helicopters passed over. Occasionally police cars would stop on the road at the end of my long drive-way. No one had come on the property, as yet. But I felt the foreboding of ignorance and misuse of power more every day.

It was only a matter of time when I would possibly be faced with escaping to another time or destroying everything. I could escape and then set in motion the destruction of everything here instead. There were so many alternatives to a civilization and a way of life that for so long and subtly had been dying. For the old man there were disseminations of his published scientific writings; he had hoped some of those like him would study and also build these things.

Here in the early half of the twenty-first century, as he had once anticipated, 'The primary experiments are over. The rocket is obsolete'. It was now that others of his caliber, the misfits, the nerds, the misunderstood, would somehow take upon them his mantel. I had begun this, my own journal, to offer to anyone who would read it, listen and look, as the old man did, and as I have.

It is written that 'prophets' are not understood in their own families and towns. They are born knowing they are different. Some of us are dreamers, gifted with abilities to reckon and understand so easily, that we seem to have no practical value to those around us.

The bullies are not always the obvious ones we have to deal with when we are children. They exist in corrupted government and industry as well as in Church and family. They are jealous and vicious and they will not understand, so they do not. They are more subtle at times too, and may appear physically weak, but have some advantage of power over us.

The old man had often entreated, 'Please build these things and just give me fair credit. Build them before someone takes these grand ideas away from you', to me and the few others he had talked to personally or in lecture, the many. His lectures had at times been sparse, until his ideas had gotten around some. I could understand that possibly all this might take a lot for the general public to fathom, even for some academia too. Not everyone would be in agreement with the presentations of his ideas, despite his authoritative perspective of actually having used the mechanism. Thusly not just ideologically based on assumptions, rather these were scientifically and experimentally based.

The future had some allure too. It was the unknown, but could I risk that destination? The past I found so comforting, and despite some obvious dangers, the past was known. How about the far distant future? There I could go, take a gander around, and if not to my liking, keep going or just return to venture back again. But the energy involved, extreme physics to render function appropriately in extreme geometry, is expensive, very expensive. Not just in capital, but in basic electrical power; it is electrical current that is essential to driving the magnetic disc upon which the Hypersphere properly sits and is articulated.

I had been invited to a Church service one day, and I went. I found it quite comforting and actually made me weep. The preacher had referred to Jesus as 'Lord and Saviour'. He invited anyone to receive Him that day. I had read a lone Bible tract tossed by someone before me along the sidewalk before this service. When the preacher said, "It is terrible to be in the hands of an Awesome God", I began to cry. He was right, for it is. God's hands can be as kind or severe as we need; yet 'terrible' too, with the meaning as 'gravely awe-inspiring' too. It is not a bad thing, though it is scary and wondrous all at the same time. The religious aspects of extreme physics are profound. When an individual is at the end of one's wits, life or emotions, for some reason beyond their understanding, that is when one really sees and hears.

Before I went home, I had stopped for Sunday dinner at a local diner. I was enjoying my coffee and had some meatloaf with catchup, boiled cabbage, sweet carrots and those great homemade, warm biscuits and butter. My hands unconsciously slipped into my pockets, and in one there was one of the discs that I had experimented with recently.

I looked over and saw one of the granddaughters of the old man with her husband and their little girl also having Sunday dinner after Church. I noticed that Alycia looked like my mother, who looked like her own mother, the mother of the old man. After I finished my meal I approached this sweet family and introduced myself.

"Oh my, why don't you join us for Sunday dinner?" offered Alycia as she recognized me, re-introducing me to her husband and daughter.

"Thank you so very much, but I must decline for I have just finished my meal across the room there," I responded with a smile.

The little girl saw me fumbling with the disc in one hand and I was somewhat nervous at not having put it away before my coming to their table. Then her mother and father saw the same and all were looking at me in my awkward position. I offered the disc in my hand to the little girl before them all, and with others in the restaurant watching.

"This was something your great-grandfather invented. Would you let me give it to you?" I slowly opened my hand before the little girl's wide-eyed face. For a moment no one seemed to know what to say, but the mother smiled, knowing that I had been a close friend of her grandfather.

"Yes, my dear," she said to her little daughter, "Pipa had made that many years ago. Would you like to put that some place special when we get home?"

The little girl nodded slightly, as not really knowing what it was. The father suddenly made a similar discovery, regarded me and lifted it from my hand. He looked it over, making sure it was not sharp or presenting any other danger, but quite intrigued by its observable internal mechanics.

"Hmm, very interesting. So what does it do?" he questioned, now more fascinated than protective.

"Well ...," I said struggling to begin and find the right words while trying to sound as matter-of-fact as I could, "it measures the changing curvature of local space-time."

"Oh, that's ... nice, yes ... very interesting." He returned it to me with a puzzled look on his face and a slight smile.

No one said much more, but the mother, the old man's granddaughter knew far more about it and she beamed quite broadly.

"Ruth, what do you say?" asked the mother, using the same name as the old man's mother.

"Thank you," said Ruth, as she cautiously took the disc. I smiled and everyone returned the gesture as I bade them a good-bye and left the diner.

I left the establishment feeling a stranger in a strange land, while feeling a sweet familiarity for the mother, the old mans' granddaughter.

It was she who as a little girl had taken a 'ride' on the swing, able to journey to the Golden Cities of the Sky. She had returned to the past where she met her great-great-great grandmother as a child. She had also met her great-great-great granddaughter of the future. She already knew so much about this disc and the Hypergeometry surrounding its function.

I was getting more easily tired. I would ask myself to consider what to do with these things the old man left with me. I even found a couple more discs. Perhaps, I should give all these things to his granddaughter, Alycia. She already had understanding of how to use them, and then leave upon her wisdom the future care of these fantastic mechanisms of geometry and Hypergeometry, geometry in the extreme and more.

I wondered if his other granddaughter, Kaylie would be interested, as she had in times past attempted the mirror experiment, similar to Alice in her entry into Wonderland. She was aware of these properties of Third Level machines too, and had moved on in life as well.

For a time both of these girls had drifted far apart from their grandfather, the old man. There had been some family rift with his son and daughter-in-law, who could never understand the powerful possibilities of what he could have shared. His son so despised him that he had disowned his father. I guess he was not supposed to re-marry and be happy, and was to afford his son and daughter-in-law to what assets he had; or let them steal all they could of who he was and what he had. The old man also had to learn from Rudyard Kipling's *The Law for the Wolves*, "For the strength of the pack is the wolf, and the strength of the wolf is the pack's." And for some of us, the worst pack of wolves is in our own family or in our own community. Jesus Christ had even spoke of such.

Alone, they present a shy cowardice. But alas it is only to bait and hurt you who are the dreamers and the thinkers, who are able to see and hear, and finally begin to grasp greater things. Yet dare not share too quickly with those ready to pounce upon your genius. The old man for years had to learn to ride the difficult, but fantastic horse that God gave him. He fell off many times, but in time learned to ride it fast, unto new times and new worlds, only the very few venture.

My old friend had told me that he would see his granddaughters in the future. He would say, "I shall see you in the future." He would at times introduce his lectures and dissertations with, "Welcome to the future … and the past, and the present!" First one was caught off guard by such affirmations. But if we think about it, it is true for all of us to be able to say to many the same. For later in the coming hours or on the morrow or the next holiday, we shall.   Even as we travel in our local one second per second rate of time passage of our 'Nows', stretching from the past to the future, we all are traveling through time, as well as space.

"Pardon, my after-image", may I say. When the Hypersphere slips into its entangled state above the driving magnetic field disc, its image as a now deep red disc, slowly lifts, fades and then becomes transparent to gone. The same as when I had carried the disc in my hand near the trolley tracks. I, along with my very local time-space oscillator, reddened and vaporized away optically. The disc in my hand at first hovered above my hand, then all was as I had described.

It is only within micrometers and millimeters away in space-time that slippage occurs. Then suddenly it is light years when all that is herein that time-space is now predominately time, and journeys to other times and other worlds have become feasible. The disc is so simply parasitic to the same distant and local Hypersphere as to be quite the way to experience such with minimal need for intrigue of mechanism. Therefore also the simplest method of use for the sphere environs of affect. Is it so difficult to understand that, between one's fingers just apart millimeters is the wavelength to the light years of the stars; and in future time, and also past time, at 370,000 times the velocity of the 'surface' at light's speed (*c*)?

If you have also been given the proper match, then you should strike it, and light it, and thus set your world on fire. Not to harm or destroy, but to teach and share such grander things to a lost world that cares not to try to know. For one common little person to be so privy of those things, that government and science are spending grand amounts of money to just tease out some more bits and pieces of these same greater things, is egregious.   While the established of privilege build their exclusive towers of 'ivory', one of meager means builds upon their kitchen table without the restriction of 'not knowing what they are not supposed to be able to do or understand'. Thus this individual finds their own great journeys of time and space.

Well, it is but a matter of time, for in less than two and one half hours one can be one hundred years ago or one hundred years hence. So God's speed, my friend, and begin your personal journeys in past and future, and to the stars, at 370,000 times the speed of light!

Fin 2 of 2

Epilogue

In the deep time of the very past and the very future, to the deep space of our Galactic eco-system, we now hopefully better can perceive our when and where, our time and place. It is paramount to consider that we, so intelligent of the creation, struggle so much with our understanding of the good and the bad, the wonderful and the terrible throughout our history, and sometimes oft think of better times in our nostalgia of long ago. If we were to be then, with our present personalities, and they now, with theirs, then would the times be really so different? In details very much for sure, but as to the Law of Large Numbers I call 'Probabilistic Mechanics', then in the greater general trends, I wager not so different. The human heart remains the same, despite the relative environments of the different times of human occupation upon this world. We are all so temporary, very temporary as we call 'today' or 'these days'.

We are so myopic in our real here and now. For when suddenly we can perceive on the grander scales of time, we find that despite many apparently different social and technological icons of the times being compared, most is more alike, than not. 'Those who do not study history, tend to repeat it.' I totally agree, for if one watches the human daily dramas of so many who are not studying, such are repeating the tragedies again and again. On this observation, those most self-serving and believing the illusion of being 'independently free', are in reality mere duplicates of the general trend. Where is the great 'free will' so often in fervor claimed? It is but for some of their misguided decisions that allows their responsibility in the greater course of the deterministic flow on far larger scales of time and space, to remain consequential. They are still very responsible though for their given measure of their decision's consequence. Do they not boast of their fortunes and victories? Yet so easily complain of their misadventures and calamities, so often brought on by their own impulsiveness, usually without moderated restraint. This appears as repeated history for the ages, given as such theater for the story of humanity's travel through time. We look back and wish to live in the 'good old days', as I have so desired myself. But given the variables of the past and now, and the constants of the past and now, such as we humans being 'human', then we can speculate with the history of humanity, that we too would laugh and cry, wonder and doubt, dream and suffer disappointment in life and death.

To escape to another time or place in the past puts us into that position of possibilities where gain or loss is sometimes severely realized. The human conquering of space helped in our perspectives being extended from the local to the global and beyond. This is the same for time, when now we have slightly better insight of the temporal graduations in increased scales of perspectives.

And this is so for future time as well. So often many people in the latter nineteenth and early twentieth centuries were enthralled with the times of years, centuries, millennia and more into the future. How impressively human life and technology would surpass those things that supposedly we could leave behind with our brave new worlds to come. We too thought we would never more suffer a dark age or any tyrannical oppressions, but soar unto grander heights of human achievements and comforts.

As to the future, yes there is always hope; and for those daring to bear responsibility to learn and be wise in their concerns and objectives, there are victories and great things to come.

But I offer that for the most of humanity, they too shall repeat history's cycles of progression and digression, and with some trend to the positive upward attainment of better things and ways of life. But these shall not be without great price. For the few need to be courageous and persistent, whilst the many gravitate to less of their potentials. Thus conflicts and wars will surely come again. Only by the grace of God shall humankind ever go beyond their egotistical presumptions of the better progression to their illusory, utopian self-made Eden.

Even H. G. Wells had alluded to such in <u>The Time Machine</u>. It seems that without much to dissuade from his premise, in the end humankind with its abandoned own self-serving intentions, would surely degrade to two such conditions of being human. Look at our pretense for such in our modern feudalistic forms of capitalism and communism. Each is a tarnished embellishment of more reasonable design. Given time, each is more like the other, where an elite few overbears a more common many. Unless humanity changes from within, it is essentially destined to the trends shadowed in our past. It is written that 'with God all things are possible'. So many of us choose not this path, for we have assumed we have understood our condition and place in time and space so well, yet our fruit rots despite our pretensions. I say, enough time will concur that the past history will reveal the future's history.

"Welcome to the future, and the past, and now." Please remember that it is up to you, you who dare to strike and light the match that God has given you, and set your world on fire. Let this fire, reflect Him Who made you and called you to be His Own.

Perhaps you will find that He all along was the One, Whom you required for completion of His ways of grander things.

"Time's glory is to calm contending kings, to unmask falsehood, and bring truth to light, to stamp the seal of time in aged things, to wake the morn of sentinel the night, to wrong the wronger till he render right, to ruinate proud buildings with Thy hour and smear with dust their glittering golden towers."
-William Shakespeare

God's speed to you, even at 370,000 times the speed of light!

Robert B Cronkhite